Julie Chamberlain
Wilhelmsburg is our home!

Urban Studies

For Wilhelmsburg, and for the Wilhelmsburgers who so generously contributed their time and energy to this research.

Julie Chamberlain, born in 1978, is an assistant professor in urban and inner-city studies at the University of Winnipeg, Canada. She did her doctorate at York University, Canada. Her research focuses on anti-racist and decolonizing approaches to urban and community development and planning, and on how residents of stigmatized neighbourhoods in Germany and Canada experience planning processes.

Julie Chamberlain

Wilhelmsburg is our home!

Racialized Residents on Urban Development and Social Mix Planning in a Hamburg Neighbourhood

[transcript]

This research was supported in part by funding from the Social Sciences and Humanities Research Council of Canada.

Social Sciences and Humanities Research Council of Canada Conseil de recherches en sciences humaines du Canada

Canada

A dissertation submitted to the Faculty of Graduate Studies in partial fulfillment of the requirements for the degree of Doctor of Philosophy, Environmental Studies. Faculty of Environmental and Urban Change, York University, July 2020
Supervisor: Jin Haritaworn; Thesis committee: Stefan Kipfer, Ilan Kapoor, Julie Tomiak; External examiners: Sybille Münch and Teresa Abbruzzese

Bibliographic information published by the Deutsche Nationalbibliothek
The Deutsche Nationalbibliothek lists this publication in the Deutsche Nationalbibliografie; detailed bibliographic data are available in the Internet at http://dnb.d-nb.de

Cover layout: Jan Gerbach, Bielefeld
Cover illustration: "Wilhelmsburg, looking northwards from the bunker", Julie Chamberlain, 2017
Proofread: Arthur Chamberlain

Print-ISBN 978-3-8376-6387-7
PDF-ISBN 978-3-8394-6387-1
https://doi.org/10.14361/9783839463871
ISSN of series: 2747-3619
eISSN of series: 2747-3635

Contents

Acknowledgments ... 7

Introduction ... 9

1. **Theoretical Framework** ... 23

2. **Methodology and Methods** ... 45

3. **Producing the "Problem Neighbourhood"**
Development of Hamburg-Wilhelmsburg in Racial Capitalism ... 65

4. **Heimat Wilhelmsburg**
Belonging and Familiarity in the Context of Devaluation ... 97

5. **Planning Strategies to "Restructure" Hamburg-Wilhelmsburg**
Mix and Mega-Projects ... 135

6. **Mixed Feelings about Neighbourhood Change** ... 169

Conclusion ... 205

Bibliography ... 213

Acknowledgments

This book and the research it is based on were brought to life by the support, attention, interest, and care of many people. I want to extend my deep gratitude to all of them.

Firstly, to the Wilhelmsburgers who participated in this study. Your curiosity, generosity, and analysis are the foundation of this work. I dedicate it to you.

To my PhD supervisor, Jin Haritaworn, for their excellent guidance and close reading at every step of the way. You have taught me so much about what it means to be an activist scholar, and I am deeply grateful.

To my doctoral committee members Ilan Kapoor, Stefan Kipfer, and Julie Tomiak. To the external examiners Sybille Münch, Teresa Abbruzzese, and defense committee chair Abidin Kusno. Thank you so much for making every stage of this work so interesting and rewarding.

To all those in Wilhelmsburg who supported or cheered me on along the way: Claudia Roszak, Manuel Humburg, Anja Blös, Marianne Gross, Karin Heuer, Barbara Kopf, Uli Gomolzig, and Moritz Rinn. The networks, ideas, time, and energy you shared with me have enabled and enriched this research immensely.

Most especially to Michael Rothschuh, who opened his home to me and introduced me to everyone and everything he knew in and about Wilhelmsburg.

To Margret Markert at Geschichtswerkstatt Wilhelmsburg & Hafen, who answered a lot of queries over several years, some of which surely made no sense as I was figuring the research out. The work you do is so important, and has enabled the historical analysis in this book.

To neighbourhood choir Kanal und Lieder, for enriching my life and soothing my soul, welcoming me with open arms and full voices. To Café Royal Salonorchester for the soundtrack to my writing.

To Anke Strüver, for her feedback and support since 2011. To Kathrin Wildner for an early conversation that encouraged and helped me make sense of my project.

To all of the wonderful students of *Hands-on Qualitative Research Methods* at HAW Hamburg in 2018 and 2019. You reminded me never to underestimate the power of good process and well-placed questions.

To my family, Kathy, Art, Carly, and Tyler Chamberlain, and Carmen and Alasdair Rodd-Chamberlain. You support my goals, cheer me on, and make me laugh. I am forever grateful.

My dad, Art Chamberlain, edited the original dissertation and this book, which was a massive undertaking that I appreciate so much. Any mistakes have since been (re)introduced by me!

To my friends and chosen family: Elizabeth Wickwire, Andrew Jehan, Jeff Myers, Felipe Vancim, Gary Pereira, Laura McKinley, Geraldo Gonzalez, Amy Packwood, Sarah Switzer and Rishi Krishnamoorthy in Toronto; Nuria Enciso and Bryan Johnston in Montreal; Doug, Christiane, and Anna Ramsey in Brandon; Sarah Zell, Fadi Ennab, Oksana Ulisko, Angela Failler, and Lyle Ford in Winnipeg. You have seen me through the various phases of this work, the ups and downs, the fun parts, and the less fun parts. Your love and companionship mean everything.

To my dear friends and co-conspirators in Hamburg and Nordborchen: Tino Kaiser, Chris Patterson, Philipp Ahlemeyer, Mahmoud Daoud, Eva Kuschinski, Rakhat Zholdoshalieva, Birgit Hasse, Yvonne Gorniak, Björn Gorniak, Enya and Janne Gorniak, and Werner and Elisabeth Bleischwitz. You welcomed me, held me up, and brought joy to everyday life while I was in Germany doing this research. Thank you so very much.

To all of my buddies from York University: Sonja Killoran-McKibbin, Amanda di Battista, Kasim Tirmizey, Ijeoma Ekoh, Ronak Ghorbani Nejad, Dena Farsad, Alexandra Gelis, Silvia Vazquez Olguin, Sheila Htoo, Stuart Schussler, Aaron Saad, Japji Anna Bas, Julia Fursova, Oded Haas, and Murat Üçoğlu.

This research and the writing of it were supported by a Social Sciences and Humanities Research Council of Canada (SSHRC) Doctoral Scholarship and travel grant, a Deutscher Akademischer Austauschdienst (DAAD) German Studies Research Grant, the research cost fund of York University Faculty of Graduate Studies, and research start-up funds from the University of Winnipeg. We should all have access to the resources that we need to produce knowledge in all of its forms.

Lastly, while I was editing the doctoral dissertation on which this book is based, nine people were killed in Hanau, Germany. They were targeted as racialized people and migrants, while they sat with friends in a shisha bar. I want to remember them, and all those who are killed by racism. Their names are: Mercedes Kierpacz, Ferhat Unvar, Gökhan Gültekin, Hamza Kurtović, Said Nesar Hashemi, Sedat Gürbüz, Kalojan Velkov, Vili Viorel Păun, and Fatih Saraçoğlu.

Unless otherwise noted, all photos and translations from German are by the author.

Introduction

Locating Wilhelmsburg

Hamburg-Wilhelmsburg has seen some major new developments over the past 15 years. The historically working class and immigrant neighbourhood in the south of the city-state of Hamburg, Germany, had struggled for decades with intense stigmatization. Headlines about the island as anything other than a space of "poverty, unemployment, and foreigners" (Gipp, 2001, translation by author) had been few and far between for quite some time. In recent years, the image has diversified somewhat, so that alongside stories about poverty, unemployment, and foreigners, there is news of music festivals, environmental organizing, basketball games at the new arena, and debate about the transformation of old factories into centres for artistic production (cf. Binde, 2019; Riebe & Röer, 2019). Where Wilhelmsburg had barely been on Hamburg's map, both literally and figuratively (Schultz & Sieweke, 2008), it now features in tourist guidebooks as a place that is well worth a visit.

This shift has been both feared and celebrated as a process of gentrification, as has also been reflected by the headlines. Articles in local and national media outlets have asked whether gentrification is taking place in Wilhelmsburg and have debated whether it is good or bad for the island (cf. Greven, 2014, 2015; Yenirce, 2017a). The media flurry has been accompanied by a deluge of scholarly research on the neighbourhood, authored by students and professors seemingly from every possible academic discipline including architecture, geography, sociology, social work, urban studies, planning, and design (cf. Birke, 2013; Christmann, 2013; Dörfler, 2014; Eckardt, 2017; Hohenstatt, 2017; Lutolli, 2021; Rinn, 2018; Schaefer, 2013b; Schlünzen & Linde, 2014; Schmidt, 2012; Vogelpohl & Buchholz, 2017; Wiedemann, 2018; Wildner & Holtz, 2015).

When I first visited Wilhelmsburg in 2011, I did not know that I would contribute to the deluge. As part of a group of Canadian graduate students on a study tour of Hamburg, I noticed that the island was a hot topic. Two massive redevelopment projects cum events had already broken ground at that time, the International Building Exhibition and International Garden Show, and we were treated to walking tours and presentations in and about Wilhelmsburg. What struck me most was how, in a talk about Hamburg's urban development and renewal strategy and priorities, the then head of planning, Oberbaudirektor Jörn Walter, told our study group that in order to solve Wilhelmsburg's problems, the population needed to be transformed. I walked away wondering "who precisely is this 'population'? And *why* did they need to be 'transformed'?" I then walked into a month-long internship at a community organization in Wilhelmsburg, and there began to find unsettling answers to my questions.

I found that the racialization of Wilhelmsburg and its residents, which is to say the production of its residents as racial "others," and the naturalization of the island as their place, were central to how they were depicted and problematized, particularly by the biggest redevelopment project at the time, the IBA Hamburg. In my master's study, I used Foucauldian discourse analysis to analyze IBA Hamburg's books, websites, pamphlets, and videos, looking at how the space and its residents were represented. I argued that the racialization of people and place, and the framing of them as in need of integration into the metropolis of the future, provided a basis on which to legitimize development interventions (Chamberlain, 2012, 2013). I later also looked more closely at how the island was framed as a "laboratory" and the IBA Hamburg as an "experiment" to justify a planning strategy that represents a risk specifically to racialized residents (Chamberlain, 2020). That research remains, to my knowledge, the only research that addresses, directly and in detail, the role of racism in the redevelopment of Wilhelmsburg.

While the image of Hamburg as Germany's so-called "Gateway to the World" is one of cosmopolitanism and openness, the image obscures how racism and the control of racialized people are part of Hamburg's landscape past and present (Della et al., 2018; Schepers, 2018). Hamburg is the country's second-largest city and its most significant port, and the structuring of the city around the production of wealth from the arrival and departure of goods and people has been central to Hamburg for essentially as long as it has existed (MacFarlane, 2019; Meyer-Lenz, 2016, 2018). The oft-celebrated diversity of Hamburg as a port city is long-standing, but has always involved differential rights, and to greater or lesser extents the control of racialized "others" in and through city space (MacFarlane & Mitchell, 2019).

Black and People of Colour scholars and activists in Hamburg have argued for decades that this is the case, and that the city's cosmopolitan image obscures the colonial logics and legacies that continue to shape urban space and development in the city-state. Profits from colonial extraction, colonial violence, and slavery enabled Hamburg's growth into the wealthiest city in Germany, with its highest per capita income and greatest number of millionaires by income (Handelskammer Hamburg, n.d.; Seukwa and Della, in Schepers, 2018). The central logic of colonialism, which "stands fundamentally for some people being worth more than others," continues to be reflected in city space through monuments and naming practices, and through the treatment of racialized people by the state and its institutions (Adjei, in Della et al., 2018, translation by author; *Afrika-Hamburg.de*, n.d.; Hengari et al., 2018; Mancheno, 2016). This has also been noted by Black German scholars and German scholars of Colour writing about Berlin, Amsterdam, Paris, and other European cities, and has led to a call to take colonialism and racism seriously in urban research (El-Tayeb, 2011, 2012; N. Ha, 2014a, 2017; Haritaworn, 2015).

Recent events on Wilhelmsburg's doorstep illustrate the urgency of this call. In December 2017 a neo-Nazi placed a bomb in a shopping bag filled with screws on the platform of Veddel S-Bahn station. The station is a busy gateway both to Wilhelmsburg and to Veddel, its similarly racialized and stigmatized neighbour. It was sheer luck that just one person was injured in the bombing and that no one else was close by enough be killed or seriously injured (Prozessbeobachtung zum rechten Terroranschlag auf der Veddel, 2018). Police and media coverage trivialized the attack, however, as "just a small explosion" that had nothing to do with the specific location where it happened (Kirsche, 2018, translation by author). The attacker, who had been convicted in the 1990s of killing a man for insulting Hitler, was characterized as a *former* neo-Nazi and a "drinker" with no apparent political motives (Hamburger Abendblatt, 2017). Not until he was convicted of attempted murder and bodily harm was the attack characterized in more serious terms as likely "xenophobic" and committed by "a fervent supporter of Hitler" (Prozessbeobachtung zum rechten Terroranschlag auf der Veddel, 2018, translation by author).

Direct, violent attacks committed by neo-Nazis have long been the *only* form of racism that were legible as such in Germany (Barskanmaz, 2011, 2012). Yet this attack was still trivialized and minimized, in a prime example of the willful logical acrobatics used by the state and by mainstream media to avoid connecting the dots when it comes to racist violence in the supposedly progressive city of Hamburg (Mobiles Beratungsteams gegen Rechstextremismus, 2018). While this kind of attack is relatively rare in Hamburg in comparison to the country as a whole, which is seeing an explosion of racist and anti-migrant violence and failing to adequately address it (Amnesty International, 2016), it raises disturbing and urgent questions for the city and thus for this book. As the flyer for a demonstration that was held by local anti-fascist and anti-racist organizers shortly after the Veddel bombing asked: "Haven't you learned anything?!" (Figure 1). If a convicted neo-Nazi sets off a bomb in a well-known immigrant neighbourhood and it is not understood as a racist attack, what hope is there that the state and media are substantially engaged in identifying and dismantling the more insidious, everyday, structural, and systemic forms of racism that undermine the well-being of racialized people in the city?

The global and local events that unfolded as I prepared to defend the doctoral dissertation on which this book is based further increased the urgency of these questions. In February 2020, a mass shooter targeted racialized people in two shisha bars in Hanau, Germany, killing nine, plus himself and his mother. From the killer's writing as well as the targets, the racist motivation of the mass murder was made abundantly clear, and even the Chancellor was moved to call out "the poison of racism" in Germany (Connolly & Oltermann, 2020; Sanyal, 2020). In April, a man who attacked a synagogue in Halle during Yom Kippur in 2019 was charged with double murder and the attempted murder of 68 people (Deutsche Welle, 2020a). In the same time period, the COVID-19 pandemic pushed deep inequalities into the spotlight, illuminating how vulnerability to surveillance, infection, and death is shaped by how people are differentially valued and positioned in societal structures and systems. High rates of Coronavirus infection in Germany's slaughterhouses and meat packing factories, for example, illuminated the miserable, exploitative working and living conditions faced by the mainly Eastern European workers in the industry, leading to outcry and promises of reform (Deutsche Welle, 2020b; Nack, 2020). Yet they also highlighted the quickness with which the public will associate "looking like an outsider" with a risk to their health, and that local officials knew about the unsafe housing and working conditions for years and took no action (Soric, 2020).

During the Coronavirus pandemic, widespread uprising against systemic anti-Black racism also arose in the United States and has echoed around the world. Sparked by the murder of George Floyd by a police officer in Minneapolis, protests against racist police violence and systemic inaction in the face of persistent racial discrimination and oppression grew and multiplied. Germany has seen massive demonstrations held in solidarity with Black people in the US, and in protest of institutionalized anti-Black racism and police violence in Germany. Both in Hamburg and Berlin, Black and People of Colour protestors reported being targeted by police, and harassed and detained immediately after demonstrations (Black Lives Matter Berlin, 2020; Neuber, 2020). In Hamburg, bystanders described how the police attacked *en masse* without warning, kettling protesters and taking 36 young people into custody." The Hamburg police justified the operation based on violations of the distancing rules that apply during the Corona pandemic. Observers and those affected speak of racist police violence: all detainees had a 'migration background'" (Neuber, 2020, translation by author). Activists against racial profiling in the city have noted that police profiling tactics overlap with COVID-19 regulations to intensify the targeting and sanction of Black and racialized people in public space (Copwatch HH, n.d.). In June 2020, a non-profit coalition also published a list of the 159 known cases in which racialized people have died in police custody in Germany over the last 20 years (Weiermann, 2020). Systemic racism and institutional racist violence have thus been prominent in public debate as I prepared this book for publication.

This research takes its cue from the scholarship and activism of Black people and People of Colour in Hamburg and in Germany, and picks up where my master's research left off, to ask what the effects of planning that is steeped in the logic of racialization are for the residents of Wilhelmsburg. I framed the study as an intersectional and anti-racist public ethnography, with the explicit goal of challenging racism in its various forms while investigating and communicating the typically marginalized perspectives of racialized people to academic and non-academic audiences (Bailey, 2013; Crenshaw, 1991; Dei, 2005; Hunting, n.d.; Tedlock, 2005). From 2016-2018 I conducted a qualitative study in Wilhelmsburg, using semi-structured interviews, participant sensing, and photography to explore how racialized residents of Wilhelmsburg experienced and engaged with recent planning and development interventions by the city-state, and how those residents figured in local planners' and politicians' decision-making. To produce this book, I brought together the resulting textual, visual, and sensory data with extensive readings of local historical, government, and media archives.

At times I have thought of this study as a qualitative evaluation of city-state planning where none was planned by the city-state itself. In response to public outcry about the prospect of gentrification, consultants conducted an annual "structural monitoring" in the last years of the IBA Hamburg, but that ended in 2013 (see Analyse & Konzepte, 2013). Today the only form of evaluation comes through the reading of census data and the city-wide 'social monitoring' report that Hamburg produces with the help of researchers at Hafen City University. I have come to understand, and this dissertation will demonstrate, that the city is intent on "mixing" the population of Wilhelmsburg as the central spatial solution to social and economic problems, and from that perspective it suffices to track demographics as the sole measure of neighbourhood change.

Wilhelmsburgers, in contrast, often framed my study as a snapshot of a moment in time – a moment in which, from their perspectives, a change had undoubtedly begun, but in which it was not yet clear where it would lead, and what it would ultimately mean for the lives of long-time residents. The fundamental changes would play out over a period of decades rather than years, and thus it was suggested to me more than once that this should be a series of studies, and that I should come back to the island in another five, 10, or 20 years, to see what had happened.

Figure 1: "Haven't you learned anything?! Fight right-wing terror and institutional racism!" In this flyer for a demonstration at S-Bahn Veddel local activists name the racism that police and media did not

(found item, translation by author, 2018)

Goals of this book

The contribution that I aim to make with this book is to the body of knowledge about Wilhelmsburg, about development and planning in German cities, and about how racialization and racism function and affect the people that they target in this present moment. I see four main parts to this contribution. Firstly, I analyze the development of Hamburg-Wilhelmsburg within the framework of racial capitalism as it connects in particular to Hamburg's colonial past and present and has produced racialized devaluation and environmental racism in Wilhelmsburg. This conceptual framework emerged from Cedric Robinson's (1983) work *Black Marxism* and from Black and People of Colour environmental justice organizing and scholarship (Bullard, 2018; Gosine & Teelucksingh, 2008), and has been elaborated by critical race (Melamed, 2015) and geography scholars (Pulido, 2016, 2017). It is uncommon to the German urban literature (for a partial exception, see Raddatz & Mennis, 2013). I offer an articulation of the local specificity of particular racializations and racisms, as Stuart Hall (1986) advocated, and connect them to broader patterns and systems that span the globe (Roy, 2018). I respond to the shortage of race-critical work in urban research about Germany, and write in solidarity with the struggles of racialized people and migrants in Germany for self-definition in the context of racialized citizenship.

Secondly, this book contributes an analysis of the city-state's planning policies and development projects in Wilhelmsburg as driven by the racial common sense of social mixing. This provides a critical addition to the local research and activism that often acknowledges differential targeting for displacement over time, but does not address the specifics of how that displaceability is produced. I bring analysis of racialization in Germany together with local research, residents' experiences, and the better- and lesser-known aspects of city-state interventions on the island, to demonstrate the centrality of the racialization of "migrants" to mixing and thus how "displaceability" (Yiftachel, n.d.) is racialized in this context. This also contributes to the ongoing scholarly critique of the logic of social mix in German planning.

This book also contributes a complex picture of how racialized people who are treated as displaceable in the logic of social mix conceptualize and rearticulate mix as an illusion and as a potentially useful tool for the community. The residents I interviewed viewed social mix in ways that both challenged and accepted hegemonic understandings of it, notably making space within the notion to advocate for the (re)integration of white Germans into spaces that they have fled and avoided. These are the 'mixed feelings' to which the title of one chapter refers: conflicting emotions and feelings in which residents support the prevailing notion of mixing, while also rejecting many of the premises on which it is based. The title draws from Jin Haritaworn's (2012) use of the phrase in *Biopolitics of Mixing*, in which they approach mixing and feelings about it from the perspectives of people with part-Thai heritage in Britain and Germany. Their work informs my understanding of social mix, and of the tension in Wilhelmsburgers' analyses, in which the problems with which they are most concerned are not perceptibly solved by the logic of mix, but rather are exacerbated by it. I argue that this presents a challenge to residents and resident activism. The book offers a potential point for critical reflection both for Wilhelmsburg residents and for politicians and planners in Hamburg beyond. It asks why planned mixing is not only tolerated but embraced, when the evidence on it is conflicting and debated, and when it plays upon exactly the kind of discrimination to which Wilhelmsburgers object in other aspects of their lives.

Finally, this book documents and communicates counter-narratives about Wilhelmsburg and contributes a valuation of the space and its residents that is different from and that subverts the dominant discourse. For research participants, this was one of the main things that my study could offer the neighbourhood, in the spirit of reciprocity that I will describe in Chapter 2. I aim to share stories and perspectives that are normally missing or marginalized in urban planning and in urban research. Residents painted a powerful picture of a close-knit neighbourhood that accepts people who are excluded from Germanness and that offers a deep sense of belonging and well-being for long-time residents. This valuation both resists and has been forged by the racism that has defined Wilhelmsburg as a "ghetto" and as a "problem neighbourhood" for as long as many people can remember. This represents a strength and potential basis for resident organizing, in the context of ongoing neighbourhood change.

Overview of the book

The organization of this book reflects what racialized Wilhelmsburgers told me were the most important themes and developments in the neighbourhood. It flows according to the priorities as residents identified them, in particular narrativizing the history of the island according to their analysis rather than simply reproducing the dominant story about 'how it came to be how it is.' As a result, the historical and development context are located in both Chapter 3 and Chapter 5, both of which draw extensively on my archival research as well as on my interviews in the neighbourhood, findings from which are woven throughout the book. This results in an unconventional structure for an academic text, as I ask the reader to first consider the meaning of the neighbourhood to long-time residents before I introduce the policies and developments that drew my attention to Wilhelmsburg in the first place. With this organizational choice I assert, as residents did, that the former is important context for understanding the latter.

The book is organized as follows. In Chapter 1, I outline the theoretical and conceptual framework that guided the study and my analysis. I anchor the anti-racist commitment that I have touched on in this introduction in a citational practice that centres the work of critical Black German scholars and German scholars of Colour. Starting from the insight that cities are produced by and themselves reproduce the structures and relations of which they are a part, I sketch out key discourses and structures of "race" and racism in the German context. I outline how I understand the process of racialization and what I see as key concepts for understanding my study and its findings, including the notions of migration background, migrantization and externalization, and integration. I then define racial capitalism as a global system that has local significance in the devaluation of racialized people, and I link racial capitalism to environmental racism and white supremacy. I argue that there are various spatializations of race and racism that prove relevant to Hamburg-Wilhelmsburg, including segregation, "the ghetto," racialized displaceability, and social mixing. These spatializations share an interest in the control of racialized people in and through urban space, whether through containment or dispersal.

In Chapter 2, I describe in more detail how I came to the project, answering a local journalist's question about how a Canadian ended up researching Wilhelmsburg. I describe how I, as a straight, white cis-woman who was born and raised in Toronto became aware of and uncomfortable with how Wilhelmsburg was talked about in Hamburg's planning and politics. By starting from how I came to the research, I situate the study as an anti-racist, intersectional public ethnography that explores racialized residents' experiences and analyses of recent developments in Hamburg-Wilhelmsburg, and how racialized people figure in local planning. I articulate the study as the production of knowledge through my embodied experience as an outsider researcher (ethnography), starting from an explicit critique with the intention to reach multiple audiences (public). I describe how anti-racist and intersectional methodologies inform the project's structure and goals, and how my methods and research practices, particularly of accountability, reciprocity, and disrupting relations of power and authority within the research itself, reflect these methodologies. I conclude by describing the methods that I used: ethnographic interviews, participant sensing, and photography, complemented by an interim report to participants and a community event that used visual and textual tools to prompt discussion. I outline who I spoke to in particular, and reflect on some of the opportunities and challenges that my methods presented.

This is followed by four data-based chapters. In Chapter 3 I trace out the negative image that has been so central to recent attempts to redevelop and revalue Wilhelmsburg, drawing on the full breadth of data from this study, including government documents, secondary historical sources, participant sensing and photography, as well as interviews with residents, planners and politicians. Wilhelmsburgers trace the island's "bad reputation" to a combination of racist discourse about the presence of migrants and material conditions in which people of low socio-economic status were deliberately concentrated on the island. I argue that Wilhelmsburg has long been approached from the outside as not-quite-Germany, in keeping with ghetto discourse and the racialization of space. The long-term process of devaluation of the island, underscored by the planning of it as a space for work and waste, is consistent with the normal functioning of racial capitalism and its production of environmental racism. Chapter 3 explores this history within the context of Hamburg as a colonial metropolis, and thus as a city with long-standing investments in racialized devaluations.

Chapter 4 focuses on Wilhelmsburgers' perceptions of their island home and in particular their appreciation of it as a space of beauty, warmth, and acceptance in the context of societal exclusion and racism. Drawing primarily on interviews with residents, I explore a side of the neighbourhood that Wilhelmsburgers argue is rarely perceived from the outside, including residents' assessments of it as a space of conviviality, where cohabitation and engagement across difference makes multiculture a part of ordinary life. Racialized residents specifically call Wilhelmsburg *Heimat*, a concept that links belonging, identity, and space. I sketch out a genealogy of the concept of *Heimat* as fraught and contested in the German-language context, where it has often been taken up in racist and exclusive ways, and where German scholars and journalists of Colour have argued for reshaping or abandoning the concept. Wilhelmsburgers use *Heimat* to assert a sense of being at home in the neighbourhood as immigrants and racialized people. Residents' sense of belonging and attachment demonstrates that there are multiple, alternate valuations of Wilhelmsburg in the context of racial capitalism.

I see these valuations as threatened by the logic of "social mixing" that prevails in the city-state's strategies to redevelop the island. In Chapter 5 I build on the context established thus far to explore the recent planning interventions that research participants suggested had been most important and impactful in Wilhelmsburg. Drawing mainly on government archives, I outline a subsidy for student housing, social mix policy in public housing, and twin event-projects – the International Garden Show and International Building Exhibition – that Hamburg has implemented since 2005. The strategies share a central goal of social mixing, through a shift in outside perceptions of Wilhelmsburg, attraction of middle class (white) Germans to the island and restriction of access to housing for racialized people. The content of the strategies demonstrates that the common sense of social mixing is race- as well as class-based, though this has been largely missed in research on the island to date. I draw on the work of local scholars and activists, media archives, interviews, and participant sensing to highlight protest and debate and in particular the critique of the event-projects as tools of state-led gentrification in Wilhelmsburg. I find that the analytical frame of gentrification does not fully capture the racialization of displaceability in this instance, where racialized people and their attachment to Wilhelmsburg as *Heimat* is devalued. The policies of Hamburg city-state, underpinned by white supremacy that is built into an exception to German equality laws, threaten the processes of emplacement, attachment, and spatialized identification of racialized Wilhelmsburgers, putting them at particular risk of unhoming.

Yet the Wilhelmsburgers I interviewed were hopeful about the prospects for continued and increased conviviality through the mixing of more white Germans into the neighbourhood. From their perspectives, mixing is desirable if it challenges the long-standing stigmatization and marginalization of the island and produces greater conviviality and connection across difference. In Chapter 6 I draw on interviews, participant sensing, and media archives to explore these perspectives as being in tension and conflictual, even as they open space around the common sense of mixing to see it another way. I demonstrate that many of the recent changes that residents describe in the neighbourhood are creating hardship and increased vulnerability specifically for racialized people and people with low incomes. Housing has become more expensive, scarce, and less accessible for people with "migrant-sounding names," which traces back to the planning and development strategies that I describe in Chapter 5. Residents' accounts of space-related tensions and contestations between different groups on the island also underscore that mix is not necessarily leading to the kind of close living-together rather than alongside one

Figure 2: Welcome to Wilhelmsburg: Looking westward along the dike (Klütjenfelder Hauptdeich) with Harburger Chaussee to the left and the Spreehafen to the right

(photo by author, 2016)

another that some Wilhelmsburgers said they long for. In the Conclusion, I identify some of the questions that remain open from this study, particularly around the future of conviviality in the changing neighbourhood. I summarize the ground that the book covered and identify several areas for future research.

Figure 3: Map of Hamburg showing Wilhelmsburg and other parts of the city that I mention in the following chapters

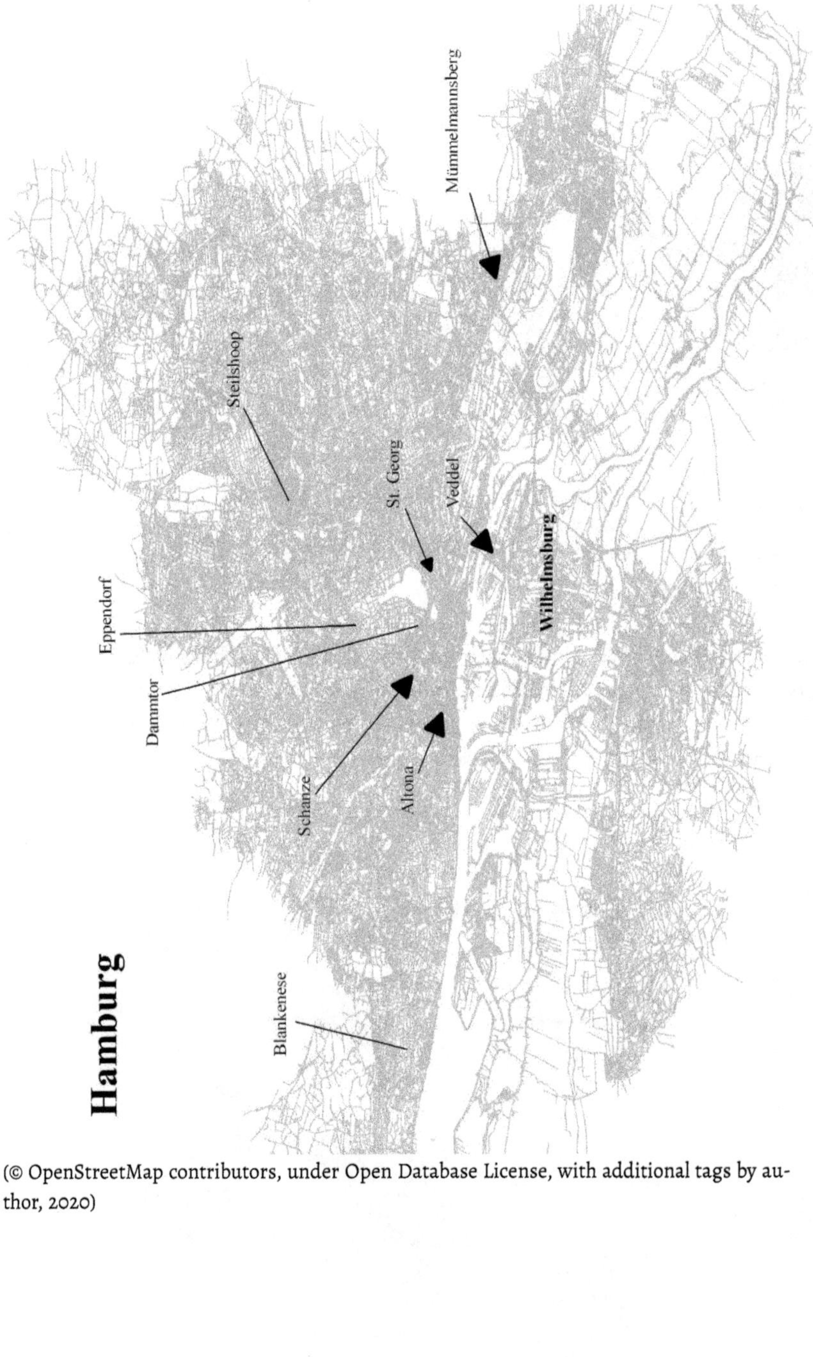

(© OpenStreetMap contributors, under Open Database License, with additional tags by author, 2020)

Figure 4: Aerial view of Hamburg-Wilhelmsburg. To see a companion street map with tagged locations mentioned in the book, go to https://juliechamberlain.org/map-of-wilhelmsburg

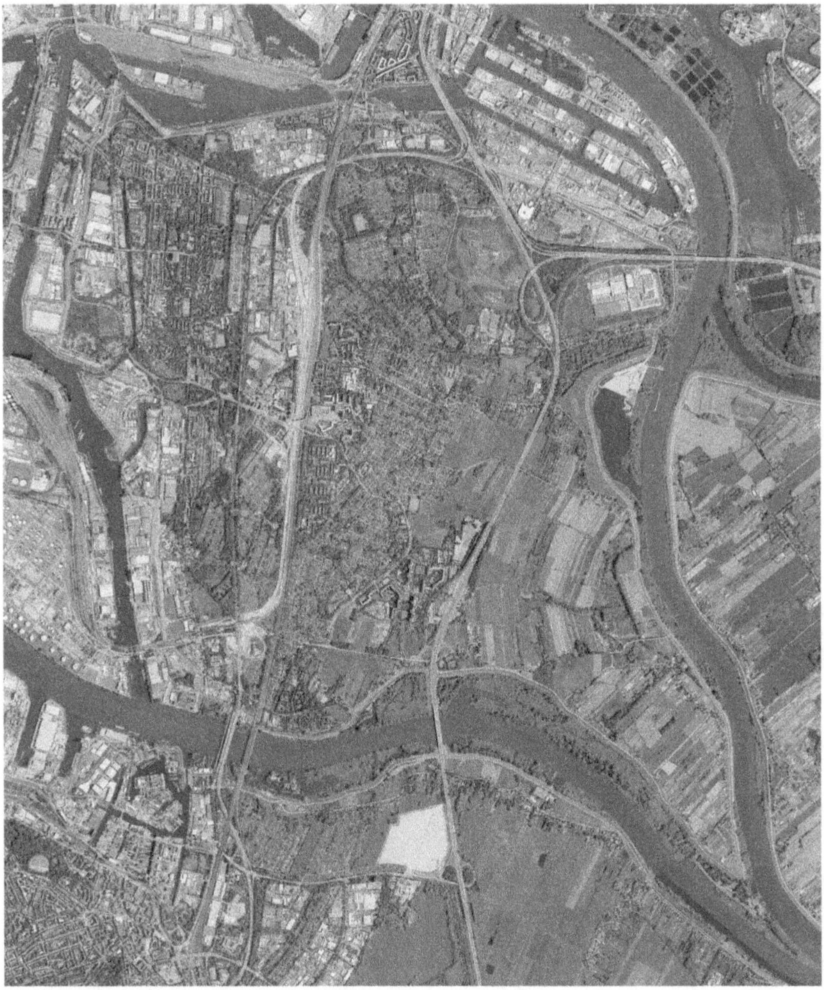

(Freie und Hansestadt Hamburg, Landesbetrieb Geoinformation und Vermessung, 2022)

1. Theoretical Framework

Introduction

In this book, I take an interdisciplinary approach to the investigation of racialized residents' experiences of development in Wilhelmsburg, drawing on scholarship from critical race and ethnic studies, urban studies, geography, sociology, and cultural studies. Two main threads weave throughout: an analysis of how racialization functions specifically in the German context, and of the articulation of racialization and racism in and through urban planning and urban space.

My approach is informed by the notion that cities exist in a symbiotic relationship with the systems of which they are a part (King, 1990). They are embedded in social, economic, and political structures and relations, and they also help to produce and reproduce those structures and relations (King, 1990; Tomiak, 2011). In this chapter, I will establish that racialization and racism are crucial structures and relations in the German context, as the very notion of Germanness in dominant discourse and practice is dependent upon racialization and externalization of Others. This, in turn, is spatialized in cities through forms of segregation and containment, and through the production of racialized displaceability, which reproduce dominant ideas of who belongs in the city and in the nation. I theorize these through the lens of racial capitalism, a concept that describes the devaluation of racialized people as central to the prevailing economic system (Robinson, 1983), and racism thus as systematic, institutional and interpersonal in nature and manifestation (Pulido, 2016, 2017).

Black German scholars and German scholars of Colour have demonstrated that European cities are spaces of colonial encounter that are invested in the control of racialized bodies in and through city space (N. Ha, 2014a, 2017), and have identified European cities, as spaces with visible racialized populations, as an important "battleground" for struggles over belonging, culture, and values (El-Tayeb, 2011, 2012). Yet the body of urban scholarship that takes racism in Germany seriously, and that directly addresses the role of race in the city, remains relatively small and comprised almost entirely of the work of scholars who identify as Black and as People of Colour (El-Tayeb, 2011; K. N. Ha, 2005; N. Ha, 2014a, 2014b, 2016; Haritaworn, 2015; Mancheno, 2016; Tsianos, 2013; MacFarlane & Mitchell, 2019). The dominant tendency in urban scholarship in the German context is the migrantization of racialized urban residents and spaces, which is to say the labelling of people and communities as migrants, regardless of whether they are or not. I will demonstrate that this is consistent with how racism functions in Germany in general (El-Tayeb, 2016; Sow, 2018). In this context even critical urban research often respects the long-standing taboo around mentioning "race," and thus struggles to make sense of the experiences of marginalized and stigmatized people and neighbourhoods. This avoidance of racism reproduces and naturalizes racism as a part of everyday German life and limits the analytical potential of urban scholarship.

As part of my anti-racist methodology, which I describe in detail in the following chapter, my theoretical framework draws first on the work of Black German and German People of Colour scholars, and puts that literature in conversation with work that has emerged from other, related contexts, again privileging the work of Black, Indigenous, and People of Colour scholars. Anti-racist methodology aims to "identify, challenge, and change the values, structures, and behaviours that perpetuate systemic racism and other forms of societal oppression" in *all* aspects of research, including who and what is considered authoritative theory (Dei, 2005, p. 3). The citational politics (Ahmed, 2013) of this book thus privilege critical, intersectional anti-racist theory and knowledge that has been produced by racialized people not only as experts on lived experience, but as authoritative theorizers and interpreters of the relevant issues and structures. One of the aims of this book is to resist the marginalization that takes place in ongoing attempts to reproduce academia as a white, male, colonial domain (see Ahmed, 2013; El-Tayeb, 2016).

In this sense it is also informed by Ananya Roy's (2018) argument for the production of urban research that keeps histories of colonialism, enslavement, and imperialism in view. Roy advocates for theorizing cities "from the South," where the South is not a location, but "a structural relation of space, power and knowledge, produced and maintained in the crucible of racial capitalism on a global scale" (Roy, 2018). To see from the South is to see European cities differently, to be able to pinpoint how "difference produced and spatialized in the context of colonialisms is constitutive of urban political economy" (Roy, 2018).

I begin by exploring the locally specific characteristics of "race" and racism in Germany. Stuart Hall (1986) argued that there are multiple, context-specific *racisms* that are just as important to attend to as are global systems. In this chapter I discuss both, starting from the supposed taboo around talking about race in Germany, the racisms that are framed in cultural and religious terms, and the notion of "migration background." I draw centrally on the work of Fatima El-Tayeb, which illuminates how German identity and citizenship are dependent upon racialization. In the second section, I define racial capitalism as a global system that requires racialization for the production of value, and then discuss key urban spatializations of race including manifestations of the racialization of space, and the production of racialized displaceability. These apparently oppositional processes – one tending towards containment, and the other towards dispersal of racialized people – demonstrate the multiplicity and flexibility of ways that racism is spatialized. At the core of both, however, is the logic of control of racialized people in and through urban space; as a dominant "myth" and "mantra" of German urban planning (Holm, 2009; Münch, 2014), social mix maintains this pattern while claiming to disrupt it.

Racialization and racism in Germany

Until recently, talking about "race" and racism was taboo in the German context. For many years there was a dominant illusion that since post-war "denazification," Germany had moved on to being "post-race" (Chin et al., 2009; Goldberg, 2008). As German author Mithu Sanyal puts it, there was "no racism in Germany" when she was growing up. "In the 1980s every child learned at school that race was a construct that fascists used to justify segregating and killing people. So, if race didn't exist, it naturally followed that racism didn't either. If you wanted to talk about it people looked at you as if you were the Nazi" (Sanyal, 2019b). When Black Germans and German People of Colour spoke out about racism, concepts like *Fremdenfeindlichkeit* and *Ausländerfeindlichkeit* (xenophobia, or fear of foreigners) were introduced as alternatives, which reproduced a dominant notion that the "others" of Germany were not racialized others but foreigners (Barskanmaz, 2012; El-Tayeb, 2016; Sanyal, 2019b). This represented a barrier to anti-racist action, but was also a crucial way in which racism in Germany was entrenched through the dismissal of the perspectives of people who experienced racism in their everyday life.

What pervaded for some time was what Cengiz Barskanmaz (2012) has called "German exceptionalism:" a provincial notion of racism in Germany as exclusively the most vicious, exceptional, and incomparable forms of racism. This exceptionalism dramatically limited the possibility for Germans, re-inscribed as always white through the special definition of "our" kind of racism, to perceive and address racisms in their various and structural forms as actually experienced by racialized people in the country (Barskanmaz, 2012). According to the logic of exceptionalism, only the most explicit, direct, and violent interpersonal racism was legible as racism.

My experience conducting this study suggests that this exceptionalism is still a substantial current in the thinking of white Germans. I was told more than once that racism in Germany was not the kinds of discrimination that I described in relation to my research in Wilhelmsburg, but something different. There is still an avoidance of the word racism and an official preference for referring to right-wing extremism and right-wing terror (see for example the comments of the Interior Minister after the recent shooting in Halle, in Hill, 2019). This is also related to *Machtvermeidung,* the evasion of power and particularly of race that Haritaworn (2005, p. 33) conceptualized in their work on whiteness in queer theory. Through race evasion, "relatively dominant people abdicate responsibility for their dominance," including in what are positioned as relatively progressive and critical contexts by using the socially constructed nature of "race" to deny the existence of racism (Haritaworn, 2005, p. 33, note 1, translation by author).

Yet some suggest that there has recently been a shift in the public discourse and that it is now "OK to be German and talk about race" (Sanyal, 2019b). Sanyal puts this down to the rise of the *Alternative für Deutschland* ("Alternative for Germany" or AfD) in recent years, whose often very explicitly racial and racist agenda has itself broken the taboo (Sanyal, 2019b), while also unleashing an increase in everyday racism (Ardal et al., n.d.). A growing body of research on racism has also made it harder to ignore or mis-characterize racism at official levels (Institut für Migrations- und Rassismusforschung Hamburg, 2011; cf. Aikins et al., 2021), and the frequency of racist attacks has made the taboo – and the denial wrapped up with it – unsustainable. This was demonstrated by Chancellor Angela Merkel's naming of the February 2020 mass shooting in Hanau as specifically motivated by racism (Connolly & Oltermann, 2020; Sanyal, 2020). Yet at the same time, there is also an active movement to remove the word "race" from the German constitution, an idea that has reared its head multiple times in recent history as a supposedly anti-racist step for the country (cf. Barskanmaz, 2011). Critical race legal scholars have strongly argued that this would be a step backwards, into treating the social construct and lived reality of race as unspeakable, at a moment when German institutions are just beginning to debate what anti-racist practice might look like (Barskanmaz & Samour, 2020).

In this book, I use German scholar Fatima El-Tayeb's definition of racialization to inform my analysis. In general, racialization refers to the process through which "race" is produced (Ahmed, 2002) as a socially constructed category that "once fixed, renders groups of people as inferior, thereby justifying their marginalization and exploitation" (Mirchandani et al., 2011, p. 120). Writing about the German context, El-Tayeb defines racialization more specifically as "the attribution of collective quasi-biological and/or cultural qualities that allow the perception of certain groups as not belonging, even when they are already part of society" (El-Tayeb, 2016, p. 34, translation by author). The *externalization* of "others" who have been part of German and European society for a long time, including Roma and Sinti,[1] Muslims and Jews, Black Germans, and "people with migration background," is central to this process (El-Tayeb, 2001, 2005, 2011, 2016). These externalizations are framed in cultural, ethnic, and religious terms, yet the differences that are attributed are

1 The Sinti are one of Germany's few officially recognized minorities; they are Roma people who have lived in the country for over 600 years (Sinti-Verein Hamburg e. V., 2015).

imagined to be immutable and passed down between generations, which marks them as racial rather than truly cultural or religious categories (see also Ahmed, 2002; Razack, 2008).

The notion of "migration background" (*Migrationshintergrund*) bears particular attention as it emerges in the following chapters as important to Wilhelmsburgers' experiences and central to the logic of the city-state of Hamburg's development of the island. The concept came into use in the 2005 national micro-census, and refers to people who have citizenship other than German, people who immigrated after 1950, naturalized citizens, and children who have a parent who falls into one of those categories (BAMF, 2017; Behörde für Stadtentwicklung und Umwelt et al., 2010; Destatis, 2018). Migration background is a racializing concept because it marks some people as unaligned with the nation, even though they are part of it (Ahmed, 2002). As my interviews with Wilhelmsburg residents will illustrate, one can be German born, a German citizen, self-identify as German, and yet be classified as distinct by the state based on migration background. Due to the vagaries of the German citizenship regime, this can be the case even three or more generations removed from any actual immigration, which El-Tayeb cites as an example of the often glaringly obvious contradictions that are produced by structures of racism, and which become commonplace and taken for granted (El-Tayeb, 2016). The concept was introduced after changes to German citizenship law meant that the distinctions that the state had previously drawn between "Germans" and "foreigners" became less useful as statistical instruments (Ahyoud et al., 2018; Destatis, 2018). Some people who had been so-called "guest workers," for example, which I discuss in Chapter 3, became eligible for German citizenship and thus were no longer marked as statistical "others." "Migration background" filled this gap and created a new categorization that now labels 22.5% of Germany's population, roughly half of whom are German citizens (Ahyoud et al., 2018).

The categorization is highly contested, especially by the people labelled with it, as this research will also demonstrate. While the state claims that it is intended to identify people who may be disadvantaged, scholars and activists who prefer to call themselves German People of Colour, Black Germans, New Germans, GermanPlus, or indeed simply German (cf. *Neue Deutsche Organisationen*, n.d.; Sanyal, 2019b), argue that only collecting equality data that focuses on structures of discrimination would achieve that aim (Ahyoud et al., 2018; Tank, 2017). Migration background *migrantizes* – labels some people as "migrants" who are not (El-Tayeb, 2016; Sow, 2018) – and continues the measurement of "Germanness" according to notions of German blood and ancestry (Institut für Migrations- und Rassismusforschung Hamburg, 2011; Sanyal, 2019a). In the everyday use of the concept, as well, it often simply refers to someone who is not white (Sanyal, 2019a; Sow, 2018). Mithu Sanyal refers to the results as the "3-H Formula" of Germanness. The first two Hs – hair colour and skin colour (*Haarfarbe* and *Hautfarbe*) – determine whether one is perceived as a person with migration background in everyday life. The third H – hemoglobin – determines whether one is considered a full citizen of Germany based on the persistent logic of belonging based on descent (Sanyal, 2019a).

The notion of migration background is one of the current manifestations of the long-standing production of dominant German identity in contrast to racialized others (El-Tayeb, 2016). Racialization is "one of the constants through which German identity stabilizes itself, through the exclusion of different variations of 'un-German'" (El-Tayeb, 2016, p. 35, translation by author). El-Tayeb's earlier work on the history of public attitudes towards Black Germans corroborates this, demonstrating that the dominant assumption is that Black people cannot be German and Germans are not Black, which reproduces the idea of Germanness as synonymous with whiteness (El-Tayeb, 2001, 2005). Other scholars have shown that despite 60 years of history in Germany, communities with roots in and ties to Turkey also continue to be seen as outsiders (Çalışkan, 2011; Hinze, 2013). Today the terms of exclusion are often religious rather than cultural, through a conflation of "Turkish" with "Muslim" and "foreign" (Ramm, 2010; Yildiz, 2009). Islamophobia and anti-Muslim racism are both widespread and institutionally embedded in Germany and in Europe more broadly (Arani, 2015), such that the assertion of a German "us" versus a "them" that is Muslim or *perceived to be* Muslim is typically framed not as racism but as a justified and necessary defence of the nation (Attia & Popal, 2016). This again is part of a racialization process that externalizes Muslims (and Black people and Turkish people) from belonging in Germany, though they are already part of the national community.

The continuity of this racialization and the power relations that it defines with the colonial history of Germany and Europe is made evident through scholarship that resists what German postcolonial scholar Kien Nghi Ha calls a "societal praxis of dehistoricization" (2003, p. 58, translation by author). Many of the attitudes that racialize, dehumanize, objectify, and externalize "others" from Germany today trace back to colonialism (El-Tayeb, 2005; K. N. Ha, 2003). German colonialism in Africa was informed and shaped by the racist ideology of Social Darwinism, which defined Germans as inherent rulers and Africans as closer to nature, savagery, and the animal (El-Tayeb, 2001, 2005). This legitimized colonialism for the German public, and justified brutal violence as a means of offering German development to the underdeveloped and civilization to the barbarian (El-Tayeb, 2005). In the same period, German companies imported and exploited migrant workers on a massive scale, based on discourse and policy that constructed Polish-speakers, for example, as less intelligent than Germans and as naturally suited to manual labour (K. N. Ha, 2007b). Their exploitation in dangerous, precarious, and low-paid work was framed as appropriate to the development of distinct peoples who were simply differently located on the racial hierarchy (K. N. Ha, 2003, 2007b). In Chapter 3 I will discuss in more detail and in relation to Wilhelmsburg how the underlying assumptions continue into the present day.

To make sense of anti-Muslim racism in Germany, critical race and diversity scholar Iman Attia and postcolonial scholar Mariam Popal also look to colonialism. They emphasize the role that religion played in the legitimation of colonial violence and in the production of a self-image for Europe against the "others" who had the "wrong religion" (Attia & Popal, 2016). Ruling according to a logic of Europe as an enlightened, white, and Christian power depended upon the ongoing definition of Europe in those terms (Attia & Popal, 2016), and thus on Europe as what Edward Said (1979, p. 54) called an "imaginative geography:" a mental landscape that constitutes an "us" and a "them" and boundaries between "our" territory and "their" territory that take on material significance. The "modern" version of Europe is an imagined geography that "has constantly, at different times, in different ways, and in relation to different 'others,' tried to establish what it is – its identity – by symbolically marking its difference from 'them'" (Hall, 2002, p. 60). In this way, "European" is no different than "German," it is defined in terms of what it is not. While the discourse about Muslims as non-European has intensified since 9/11 (Ramm, 2010; Razack, 2008) and since the 2000 citizenship changes made Muslims a permanent fixture in Germany (Yildiz, 2009), the boundary-making ideas themselves trace much further back in time.

One of the prominent ways in which German- and Europeanness continues to be produced and maintained against an imagined, essentialized "Islam" is through the supposed failure of Muslims to embody values of tolerance and equality, imagined as core European values (El-Tayeb, 2012; Haritaworn, 2015). El-Tayeb (2012) demonstrates this in their work on the positionality of queer people of colour in Amsterdam, as does Haritaworn (2015) in their theorization of the "queer regenerations" of racialized spaces in Berlin. Underscoring how racialization does not operate independently, but rather in "complex and shifting interactions" with other systems of oppression (El-Tayeb, 2011, p. 125), the focus of anti-Muslim discourse and practice is often particularly on Muslim men as threats to women and LGBTQ people. The former is often symbolized by the figure of a woman in a headscarf as the embodiment of patriarchal vulnerability (Yildiz, 2009). This too is an old move, as "Europe has long pointed to the inferior position of 'Oriental women' in Muslim societies as a way of asserting its own civilisational superiority" (Yildiz, 2009, p. 471; drawing on Yeğenoğlu, 1998). This book illuminates some of the effects of this assertion on Muslim women and girls, particularly through their exclusion from employment and educational opportunities in the name of German values (see Chapter 4 in particular). As Kimberlé Crenshaw established with the concept of intersectionality, people's lived experiences are wrapped up in "multiple grounds of identity" and numerous axes of oppression that include race, gender, class, ability, sexuality, and immigration status (Crenshaw, 1991, p. 1245).

While the woman in headscarf is made hyper visible through the dominant anti-Muslim discourse, El-Tayeb (2012) argues that LGBTQ Muslims disappear or are framed as impossibly exotic and too oppressed to speak for themselves. The assumption is that Muslim equates with homophobia, and thus with a threat to the safety and security of LGBTQ people (El-Tayeb, 2012; Haritaworn, 2015; Petzen, 2009; Tsianos, 2013). The framing of tolerance and equality as essential European values at the same time obscures homophobia and gender-based oppression that is perpetrated by the dominant (white German) population (El-Tayeb, 2012; Haritaworn, 2015). It also ignores and justifies increasing intolerance towards racialized people, immigrants, and people perceived to be Muslims (El-Tayeb, 2012), and justifies growing numbers of anti-Muslim and anti-immigrant attacks in Germany (Amnesty International, 2016; Arani, 2015; Pro Asyl, 2017).

In this context, the notion of integration is a powerful fiction that serves to discipline immigrants and racialized people through its unattainability. Integration requires "a national culture and universal values (possessed automatically by all 'original citizens') in which newcomers must be instructed" (Razack, 2008, pp. 129–130), and thus reproduces the definition and boundaries of the dominant society, while subjugating people to the authority of that society (K. N. Ha, 2007a). Since 2005, language and cultural instruction have been required for migrants coming from outside the European Union, which reinforces the notion that non-Europeans are fundamentally different from Europeans, and need integration education as a form of "development aid" (see Ramm, 2010), or civilizing mission (K. N. Ha, 2007a). The terms are set by the dominant population, however, which is invested in the reproduction of racialized people as always out of place and out of time (El-Tayeb, 2016).

Racialized scholars and activists in Germany have thus identified integration as a losing game and an impossibility, as I also demonstrate in Chapters Three and Four based on my interviews with Wilhelmsburgers. Muslim German scholar Lamya Kaddor (2010) argues for example that her roots and religion will always be made an issue of, no matter what she does. Referencing stereotypes of German culture, she says that even if she "[ate] sauerkraut with pork knuckles, or put on lederhosen, [sat herself] down at the bar with a glass of *Weissbier* and philosophized about the long-lost *Spielkultur* of the national football team" (Kaddor, 2010, pp. 105–106, translation by author), she would still continuously have to explain and defend her "origins" and her religion. As a result, she advocates for a complete reimagining of the notion of integration and of belonging in Germany.

The picture that this scholarship and activism in Germany paints is one in which racialization is consistent and persistent, but also variable in its specifics. This echoes the broader conceptualization of racialization as changeable and flexible over time and place (El-Tayeb, 2016; Mirchandani et al., 2011). A key aspect of this changeability is illustrated in how some groups of people, once racialized, are later incorporated into the nation (Pulido, 2006). Geographer Laura Pulido (2006, p. 24) calls for an analysis of "differential racialization" to understand the "subtle and not-so-subtle differences" between how groups are racialized in relation to specific histories, geographies, inclusions, and exclusions. Racialized people cannot be approached as a homogenous group who all face the same experiences and conditions, as differential racialization "affects how each group is treated legally, socially, and economically and can even determine life and death" (Pulido, 2006, p. 25).

Racism, then, is not just about "othering," it is about access to rights and resources and basic forms of well-being (Sanyal, 2019a). It is critical to the capitalist economic system (El-Tayeb, 2016) and creates the conditions for forms of externalization, exclusion, and devaluation that fundamentally affect racialized people's well-being and life chances (Gilmore, 2002). Further, racialization and racism are implicated in the production of urban space, as cities are both the products and reproducers of the structures and processes in which they are embedded (King, 1990). In the following section I present several concepts that are central to my analysis of the development of Wilhelmsburg and of racialized residents' experiences in the neighbourhood, as situated within structures and processes that are locally articulated but reach far beyond the local context. I discuss my understanding of racial capitalism, environmental racism, and white supremacy, which I base on critical race and geography scholarship that draws on the Black Radical Tradition and particularly on the work of Cedric Robinson. Bringing this scholarship together with the analyses of the German context in particular enables a view of Hamburg and Wilhelmsburg as embedded in global processes to which racism and racialization are central. I highlight key concepts including segregation and the ghetto, as well as their related "opposite," racialized displaceability. In the latter concept I include the discourse and practice of social mix, which depends upon racialized devaluation to function. Bringing global processes together with international and German scholarship on the racialization of urban space and production of racialized displaceability counters the innocence in relation to colonialism and slavery that is often implicitly claimed by German urban research.

Race, racial capitalism, and the city

Racial capitalism is a concept that was introduced by Cedric Robinson (1983) in *Black Marxism* to characterize how racism is crucial to the basic functioning of the economic system. Arguing that "racialism" pre-dated capitalism in the justification of labour exploitation and slavery in European history, Robinson asserted that capitalism has always been systemically *racial*. This is fundamentally distinct from a view of race and racism as "additive" to other forms of oppression and to the economic system (Pulido, 2016). An analysis of racial capitalism requires attention to the origins of power to notice how we are still living with the legacies of colonialism, imperialism, and enslavement as systems that have produced and maintained capitalist wealth and unequal patterns of development (Pulido, 2016, 2017; Roy, 2018).

Contemporary theorists of racial capitalism have elaborated on *how* it devalues people and spaces to produce value. Critical race scholar Jodi Melamed (2015) explains that in order for capital to accumulate, it must move through relations of severe inequality among humans. "Racism enshrines the inequalities that capitalism requires ... displacing the uneven life chances that are inescapably part of capitalist social relations onto fictions of differing human capacities, historically race" (Melamed, 2015, p. 77). An analysis of racial capitalism thus requires attention to how those fictions of different capacities are produced through processes of racialization.

Laura Pulido further elaborates that the required inequality and thus devaluation of racialized bodies is produced in two principal ways: through racialized systems of labour that sort bodies and produce some that are surplus, and through the appropriation and control of access to land (Pulido, 2017). These devaluations create "a landscape of differential value which can be harnessed in diverse ways to facilitate the accumulation of more power and profit than would otherwise be possible" (Pulido, 2016, p. 1). One of the ways this takes place is through the production of environmental racism, which Pulido theorizes as part of the normal, everyday functioning of racial capitalism (Pulido, 2016, 2017). Environmental racism means that historical processes have led to people who are seen as undesirable – racialized people, immigrants, people with low incomes – carrying a disproportionate amount of environmental risk and burden (Bullard, 2018). Gosine and Teelucksingh (2008, p. 45) have also defined environmental racism more broadly as the convergence of undesirable people and undesirable land uses. Situating environmental racism within racial capitalism suggests that "undesirability" and thus devaluation are in fact produced together, as a constituent aspect of the production of value (Pulido, 2017).

One of the forms of racism that enables this co-production of devaluation is white supremacy. Pulido (2015) theorizes white supremacy in relation to the long-term persistence of environmental racism, in order to make sense of the apparent failures of the state and corporations to protect the well-being of racialized and poor communities. Using the example of Exide Technologies, a battery-recycling plant in Vernon, California, Pulido identifies three aspects of white supremacy, while noting that there are likely others: *awareness* that people of colour are being harmed, *taking* from people of colour, for example by displacing the cost of doing business onto them, and *racial superiority,* as manifested for example in the apparent belief that those being harmed are expendable (Pulido, 2015). Through this analysis Pulido illuminates that there are limits to the notion that environmental racism harms racialized people unintentionally, and she advocates for attention to the specific forms of racism at play in a given situation (Pulido, 2015). In Pulido's assessment, the state is not an ally in the struggle for racial and environmental justice, but rather is implicated in the maintenance of injustice and inequality (Pulido, 2015).

Pulido's work has been at the forefront of a recent flourishing of scholarship that explores the specific interconnections of racial capitalism and cities of the global North. The framework encourages analysis beyond national borders to understand the production of urban space in relation to patterns of racial hierarchization, expropriation, and exploitation (Dorries et al., 2019; McCreary & Milligan, 2018; Roy, 2018). Approaching urban space through racial capitalism invites an analysis of "who can claim home and land" in the city (Roy, 2018) and who, in contrast, is treated as though their lives do not matter (Ranganathan, 2016). Scholars have centrally implicated liberalism in the making and maintenance of this distinction (Lowe, 2015; McCreary & Milligan, 2018; Ranganathan, 2016). Writing on the "slow poisoning" of Flint, Michigan, and in conversation with Pulido's discussion of environmental racism, Malini Ranganathan argues that liberalism, as a system of political thought, centres a white, male, individual in contrast to others who are seen as "unfamiliar and unruly" (Ranganathan, 2016, p. 6). Liberalism contains exclusion within it, and yet in discourse and practice tends to obscure how race is foundational to the economic and political order (Ranganathan, 2016). As a result, liberal forms of recognition of inequality, for example in urban environmental governance, tend to treat race as simply a marker of "vulnerability" and not as part of a system that urban governance, politics and planning themselves reproduce and perpetuate (McCreary & Milligan, 2018; Roy, 2018).

Recent analyses of cities through racial capitalism have addressed the ongoing significance of colonialism, imperialism, and slavery, to the production of urban space both in settler colonial cities and colonial metropoles. Dorries, Hugill and Tomiak (2019) analyze the roles of racial capitalism and colonialism together in the example of the city of Winnipeg, to argue that the systems share property regimes that are shaped by racial hierarchies, dehumanization, and expropriation. The lens of racial capitalism draws particular attention to how, when, and where "racialized accumulations of property and advantage" have been naturalized (Dorries et al., 2019, p. 6). In a distinct but related analysis, Ida Danewid (2019) makes sense of the Grenfell Tower fire in London through racial capitalism, and emphasizes how the city has prospered through accumulation by dispossession, and as a site through which colonialism and imperialism have been financed and managed. Danewid argues that cities of the global North continue to be "places where racialized forms of dispossession and expropriation are orchestrated and reproduced," devaluing certain people and places in a devastating and deadly fashion (Danewid, 2019, p. 16).

The differences between and among cities are interpreted in this literature through racial capitalism as "a dynamic mode of organizing the world that articulates in particular ways in different contexts" (Dorries et al., 2019, p. 2). Racial capitalism was first conceptualized in relation to Europe, and is now taken up to theorize a range of distinct and interrelated geographies. Hamburg's location within racial capitalism will necessarily be distinct from that of Winnipeg, London, Flint, or even Berlin. I will demonstrate in Chapter 3 that colonialism and slavery play crucial roles in that location as they do in other cities. While never the capital of Germany, Hamburg has been called the country's "number one colonial metropolis" (Zimmerer, in Norddeutscher Rundfunk, 2018, translation by author) because of its long-standing investments in the kinds of financing and orchestration of colonial trade and governance to which Danewid points. Profits from German colonialism and European colonialism more broadly, including from related processes and structures of slavery, extraction, and expropriation, have made Hamburg the city it is today (Seukwa, in Schepers, 2018). Recent scholarship on the city's "colonial legacy," and its official and everyday "amnesia" in relation to this legacy, demonstrates that the city's colonial foundations are most visible in its relationship to its "others," which is defined by the colonial logic of "some people being worth more than others" (Adjei, in Della et al., 2018, translation by author). The lens of racial capitalism draws our attention to how this is both an old and ongoing dynamic that continues to play out in and on urban space. Colonialism is a structure rather than an event, as Patrick Wolfe (2006) has argued in the settler colonial context, and which is reflected in the analyses of Hamburg I will further present in Chapter 3. The value of particular people and spaces are defined through racial categories that are continually in the (re)making through racialization (McCreary & Milligan, 2018). These are reproduced and operationalized in urban space to create surplus and power, sometimes in ways that even exceed the desires of capital itself (Pulido, 2017). Such is the dynamic nature of racial capitalism.

Racial capitalism informs the control of material resources, including the basic necessities of life, access to wealth, housing, and the distribution of environmental risks and benefits through what Sherene Razack (2002, p. 6) calls "the spatiality of the racial order in which we live." This spatiality includes the racialization of space, which I discuss in the following section, with a focus on segregation and "the ghetto" as examples that are relevant to Germany and to Hamburg. I then turn to spatializations of race and racism that involve dispersal and displacement of racialized people. Though the spatial strategies are distinct, the logic of control of racialized people in and through space is consistent.

Racialization of space: Segregation and the ghetto

Racialization of space is perhaps the most widely recognized urban spatiality of the racial order, and proves to be critical to understanding the past development of Wilhelmsburg from residents' perspectives. The racialization of space means that certain spaces come to be associated with particular "races" and naturalized as the correct place for those racialized bodies (James, 2012; Razack, 2002). Racialized spaces are thus examples of the imaginative geographies that I touched on above in the section on racialization in Germany: mental boundaries between "us" and "them," and "our" territory and "their" territory, that originate in the mind but take on material significance (Said, 1979). Rather than a simple spatial distinction between natural, pre-existing groups, imaginative geographies are part of the very constitution of an us and a them (Said, 1979). The racialization of space and of bodies is thus a dialectical relationship (Razack, 2002) that communicates racial meaning, while also producing it.

"The ghetto" might be considered a quintessential racialized urban space. The concept of the ghetto has historically signified that part of the city to which racialized people are confined (Stehle, 2006). It has a symbolic and material connection to segregation, which Carl Nightingale (2012) defines as the compulsory and unequal separation of groups in city space, in service of elite accumulation of power and wealth. Segregation has been practiced in Europe since at least the Middle Ages (Nightingale, 2012), and as a concept, the ghetto travelled from Europe to the United States and back again, maintaining its core meaning although the specific discourses about racial segregation changed and shifted over time and space (Stehle, 2006).

In recent years, "the ghetto" has been taken up in Europe as a cautionary racist discourse about the need to control racialized people in urban space and particularly through the neighbourhoods in which they live (Stehle, 2006). Drawing on its tropes of violence, lawlessness and social decay, references to the ghetto communicate fear and a sense of threat. In its travels the concept has picked up a particular association with Black people in the United States and with the "American-style ghetto," and anti-Black racism ensures that that association is popularly read as negative and dangerous (Rodatz, 2012; Stehle, 2006). Labelling places as ghettos is thus a powerful way of signifying race and assigning racist meaning to them, without necessarily having to say the words. At the same time, the travels of the ghetto to the U.S. and back again contribute to the myth that is present in German-language literature about the European city: that race and racialized spaces are exclusively American problems (cf. Häußermann, 2001). The ghetto is thus "constructed as a patriarchal, violent, and non-European space in the centre of Europe ... a space that is 'a problem' that 'we' need to deal with" (Stehle, 2006, p. 61).

The racialization of space reproduces social, economic, and spatial injustices while treating local conditions as though they are produced by the presence of racialized bodies themselves (James, 2012). Often moral panics figure prominently in the discourse around racialized spaces, for example in relation to violence, drugs, or sex work (Anderson, 1991; James, 2012). Moral panics produce a mandate for state action (Tsianos, 2013). They justify policing of bodies in the space and draw attention away from similar "immoralities" in whiter and wealthier communities (Haritaworn, 2015). The scrutiny and policing encouraged by moral panics fuels additional panic, scrutiny, and stigmatization, which further masks and justifies distributional and environmental injustices (James, 2012).

Discourses and fears about "self-segregation" and "self-ghettoization" are central to the interlocking of racialization and space in Germany (El-Tayeb, 2012; Haritaworn, 2015; Stehle, 2006). The dominant narrative is often about people "isolating themselves" (Drever, 2004; Hinze, 2013; MacDougall, 2011), particularly into so-called "parallel societies" (*Parallelgesellschaften*) that are imagined as within Germany but somehow beyond the control of the state (Çağlar, 2001). The "parallel society" is a concept that was introduced by sociologist Wilhelm Heitmeyer in 1997 and has since been widely used to demonize racialized people for their connections to one another and for their concentration in particular spaces (Tsianos, 2013; Rodatz, 2012).The implications go against the evidence about the extent and causes of segregation in Germany, which is in fact very minimal and caused mainly by exclusion that is perpetrated by the dominant white, German population (Harlander, 2012; Münch, 2009, 2014; Tsianos, 2013). Yet the discourses of self-segregation and self-ghettoization stoke fear of racialized communities and spaces, ignoring or concealing the discrimination in housing, work, and education that shapes communities' lives and opportunities and suggesting that "migrants," particularly people with roots in Turkey and people perceived to be Muslim, choose to segregate themselves rather than "integrate" with the dominant population (El-Tayeb, 2012; N. Ha, 2014b; Münch, 2009).

Vassilis Tsianos (2013) calls this the Integration-Ghetto-Complex. Drawing on research about the St. Georg district of Hamburg, Tsianos argues that racism is not typically addressed in urban research and policy, and that segregation and dis-integration are treated as the same thing. The state pays little to no attention to structural and institutional factors that produce so-called "problem neighbourhoods" (Tsianos, 2013); instead, the urban spatialization of race is treated as evidence of the ongoing failure of the "migrant Other" to integrate, and thus as the basis for continued racialization and exclusion, as well as state intervention (see also Chamberlain, 2012, 2013; MacDougall, 2011). The concept of the problem neighbourhood has thus been criticized as a product of "urban Orientalism," which – similar to the Ghetto – imagines the culture of an "other" that does not fit with the values of the bourgeois city, and thus evokes a non-European space (Lanz, 2002). The "problem neighbourhood" is thus also a mechanism through which a racialized ghetto discourse is mobilized into concrete planning and policy in the German context (Keller, 2015; Tsianos, 2013). It shares its central tropes with the ghetto, and thus the implication of a space that is "'a problem' that 'we' need to deal with" (Stehle, 2006, p. 61).

The Integration-Ghetto-Complex and the racialization of space more generally obscure the actual complexity, diversity, and strengths that are present in neighbourhoods labelled as so-called problem neighbourhoods (Keller, 2015; Tsianos, 2013). Spaces of exclusion can be sites where notions of integration, belonging, and identity are contested, not to mention sites of community, solidarity, resistance, and thriving economy (Çağlar, 1995, 2001; Çağlar & Schiller, 2018; El-Tayeb, 2011; Haritaworn, 2015; Hinze, 2013; Schiller & Çağlar, 2013). Indeed for people who are excluded from the dominant definition of German, the city or the neighbourhood can be the site of meaningful identification and belonging, as research has found for example in Berlin (Çağlar, 2001; Çalışkan, 2011; Hinze, 2013). Canadian sociologist Gulhanim Çalışkan's (2011) study with people she calls "German-born Turkish *Ausländer* (foreigners)" – a framing that highlights one of the absurdities created by the yoking of racialization and German identity – found that her interviewees identified fondly with the city of Berlin or with the neighbourhood of Kreuzberg, and through that identification challenged the established notion of who belongs in Germany.

One of the concepts that emerged from my interviews with Wilhelmsburgers to characterize their identification with the stigmatized neighbourhood is *Heimat*. *Heimat* is a common word in German, but does not translate directly into English; depending on the context it can be translated as home, homeland, hometown, birthplace, or as other related English terms. Because of this untranslatability into English, I have elected to leave the word in German and to italicize it throughout the book. As a spatial concept of identity and belonging (Blickle, 2002; Boa & Palfreyman, 2000; Eigler & Kugele, 2012), *Heimat* is a kind of imagined geography, but in a slightly different sense than after Said. *Heimat* references a subjective sensory and feeling relationship to space (Römhild, 2018) and is often associated with nostalgia for a place and time that is idealized and may have never really existed (Blickle, 2002; Costadura et al., 2019). It *can* be used to create and reproduce boundaries of "our space" versus "their space," which is how it was used under National Socialism and is often mobilized by the racist right-wing today. As a result, the concept is currently highly contested (cf. Ataman, 2018a, 2018b; Aydemir & Yaghoobifarah, 2019). Yet I will demonstrate a non-hegemonic use of the term by racialized Wilhelmsburgers, who use it to claim *home* in an inclusive and open fashion.

Alongside the racialization of space, which involves the fixing of bodies in place (Razack, 2002), urban spatializations of race as eviction, exclusion, and dispersal through policies and practices of "social mixing" also emerged as central to residents' analyses of neighbourhood change in Wilhelmsburg. In the following section I discuss these spatializations, framing them in terms of a racialized production of *displaceability* (Yiftachel, n.d.).

Racialized displaceability and exclusion

Geographer and urban planner Oren Yiftachel (n.d.) argues for an analytical shift away from displacement as an *act*, to displaceability as a *systemic condition*, in order to approach both as differential. Displaceability refers to "the susceptibility of people, groups and developments to be removed, expelled or prevented from exercising their right to the city." Emerging from what Yiftachel calls a Southeastern perspective, a perspective that emerges from the global South and East in contrast to the dominance of urban theory from the global North and West, the concept acknowledges that urban residents face different "depths" and "threats" of displaceability in relation to their legal, political, social, and economic status (Yiftachel, n.d.). Writing from Israel, Yiftachel argues that a Southeastern perspective takes "identity regimes" particularly seriously, as a "sphere of power where the status, resources and visibility of groups are determined, negotiated and challenged" (Yiftachel, n.d.). In this sense it is compatible with the theoretical lens I have presented thus far, in which race and racism are central to urban political economy and to the production of space.

This book is informed by the identification of multiple forms of racialized displaceability in the critical race urban literature. Urban sociologist Christopher Mele (2015), who writes about how race and exclusion are structured in American cities, argues for example that the privatization of previously public spaces in the neoliberal city contributes to the exclusion of immigrant and racialized communities from public space. The exclusivity of consumption-based spaces demands certain economic capacity to gain access and to belong, and immigrant and racialized communities are systematically denied this economic capacity (Mele, 2015). Neoliberal urban development nonetheless ensures the creation of ever more spaces that are defined in this way (Mele, 2015).

Markers of economic status are also at times used as cover for racial discrimination in urban space (May, 2014). In the anti-Muslim urbanism that Tsianos (2013) identifies in Hamburg, the city targeted certain kinds of businesses for systematic expulsion from the neighbourhood of St. Georg, in an attempt to reduce and control the Muslim and low-income population. The city encouraged restaurants to create outdoor eating areas as a means of controlling the streets and squares, and selectively restricted what kinds of businesses to which building owners could rent, in order to exclude those perceived as catering to the racialized and Muslim community (Tsianos, 2013). In this way, "older racial orders are reshaped and revitalized" (Razack, 2002, p. 16) in neoliberal terms that are superficially about the capacity to consume. While displacement may take place on the basis of failure to "achieve consumer-citizen status," as El-Tayeb (2012, p. 81) puts it, racialization is central to the displaceability of certain people over others.

Urban racialized displaceability is further manifested through racial profiling in policing. Racial profiling, particularly of Black men and people perceived as "Turkish," "Arab" or "Muslim" has been identified as an ongoing problem in Hamburg (Popp & Gezer, 2013; taz, 2018) and across Germany (UN Human Rights Council, 2010, 2017). Urban sociologist Noa Ha (2014a) has identified it as a manifestation of the persistence of colonial logics of control in European cities. Racial profiling represents a moment of colonial encounter, Ha argues, wherein the colonizer (white, German police power) and colonized (racialized "others") come into contact, and control of racialized bodies is attempted and contested (N. Ha, 2014a). In Hamburg, the city has officially labeled particular areas as "dangerous places" in order to facilitate surveillance, control, and expulsion by police. Locals to those areas, which currently include parts of St. Pauli and St. Georg, argue that the status is tied directly to practices of stopping, IDing, and dispersing racialized people (Kampagne für die Grundrechte, 2013; MacFarlane & Mitchell, 2019).

The manifestation of racialized displaceability that is most central to this research is that of "social mix." Social mix refers to the planned pursuit of a mix of urban residents at the level of neighbourhood, block, or building, typically according to income, housing tenure, and/or "ethnicity." It also refers to the discourse that such a mix is desirable. Social mix policies are implemented in many different ways. Typical strategies include subsidies to encourage the moving-in of particular kinds of tenant, as well as housing diversification, to attract people with various incomes and interests in buying or renting housing (Bolt, 2009). The "mix of thirds" (*Drittelmix*) that has been introduced in Hamburg in recent years, is an example of the latter. It requires that of any new housing built in the city, a third of the units should be owner-occupied, a third market rental, and a third publicly subsidized rental units (Freie und Hansestadt Hamburg, Senatskanzlei, n.d.). The restriction of housing allocation has also been a common strategy of social mix, particularly in Germany. German cities have, for example, used bans and quotas to restrict where people of certain "ethnic" backgrounds are allowed to live. While bans have fallen out of fashion, quotas remain common (Harlander, 2012; MacDougall, 2011; Münch, 2009), as I will discuss in particular in Chapter 5.

Prevailing theories of what social mix might accomplish in a neighbourhood are so extensive and varied that German sociologist Andrej Holm (2009) argues that there is no other concept so laden with meanings and motives in the German urban planning context. In a review of international social mix literature, geographers Bolt, Phillips, and van Kempen (2010) identify three main logics that comprise the discourse and justify the practice. They include providing housing "careers" for residents, wherein the planned creation of an income and tenure mix is thought to offer residents housing options within a neighbourhood; social contacts, wherein mixed districts increase cohesion among urban dwellers with different incomes, class, and "ethnicities;" and social capital and social mobility, in which it is thought that mixed neighbourhoods create "productive contact" between residents and discourage what are considered to be problematic bonds that limit social mobility (Bolt et al., 2010). Gentrification scholar Loretta Lees also identifies two more logics: the "money go round" logic, in which middle-class residents are central to the neighbourhood economy, and the neighbourhood defence logic, in which they are powerful advocates and thus key ingredients to any neighbourhood (Lees, 2008, drawing on Schoon, 2001). Lastly, the theory of neighbourhood effects often plays an important part in the underlying logic, as we will see is the case in Hamburg. The idea of neighbourhood effects is that "deprived neighbourhoods have an additional and independent negative effect on the well-being of their residents" (van Ham, in Housing Europe, 2015). If that is the case, then the logic asserts that mixing "deprived" neighbourhoods will improve conditions and thus the lives of all residents.

Social mix has achieved the status of common sense in German urban planning, with the discourse so ubiquitous that the desirability of mix is assumed, often without explanation (Holm, 2009; Münch, 2009, 2014). The practice has been around since at least 1863, when Hobrecht's designs for Berlin neighbourhoods centred an ideal of mixing the working and middle classes in each complex (Holm, 2009). Public housing allocation has also been informed by the ideal to greater and lesser extents for at least 100 years (Münch, 2009). Two key themes have been consistent in German social-mix discourse and practice in this time: the primacy of a bourgeois ideal of the urban resident, and the notion of breaking up and discouraging the development of "uncontrollable" neighbourhoods through mix (Holm, in Bayer et al., 2014; Holm, 2009).

These themes are consistent with Justus Uitermark's (2014) analysis of social mix as a form of social control and governmentality after Michel Foucault. Governmentality, according to Foucault, refers to the governance of populations directly and indirectly through institutions, procedures, and analyses, with techniques of which the population is not always aware (Foucault, 1991). Reading social mix through a governmentality lens, Uitermark argues that social mixing extends state governance into urban neighbourhoods in the person of middle class (white, German) families and civil society actors who "embody respect for the state," and on whom the state relies for support of the dominant order (Uitermark, 2014, p. 1430). In a context where, as this research demonstrates, social mix is often mobilized specifically in relation to "migrants," this form of governmentality is closely tied to the Integration-Ghetto-Complex. Social mix operates as a technique for the "integration" of "others" in urban space (Uitermark, 2014). As I noted above, however, the definition of Germanness through the 3-H Formula forecloses integration as a real possibility. Social mix as a form of social control thus also serves to discipline the population regardless of whether its claims are fully realized, which critical reviews of social mix policy and practice indicate is rarely, if ever, the case (Arthurson, 2012; Bolt, 2009; Bolt et al., 2010; Ruiz-Tagle, 2016; Saville-Smith et al., 2015).

Social-mix policies have been referred to as "myth" (Holm, 2009) and as "faith-based policies," because they are not well supported by evidence that they actually do what they claim (Cheshire et al., 2008). The underlying logics and assumptions of social mix do not pan out in practice. Living in a mixed neighbourhood does not necessarily produce interaction, cohesion or respect; instead mix has often been associated with increased conflict (Arthurson, 2012) and with interactions that are marked by fear, distrust, and avoidance (Ruiz-Tagle, 2016). Further, middle-class people do not necessarily share their social capital or make opportunities accessible to low-income and racialized others (Nast & Blokland, 2014). Rather, Wehrheim suggests that middle-class people tend to appropriate space, shape it in their own image, and isolate themselves from low-income neighbours (Wehrheim, 2018). As a result, social-mix experts argue that there is "no solid evidence that mixing actually benefits the original population of deprived neighbourhoods" (van Ham, in Housing Europe, 2015). Instead it can break up communities, decrease diversity, increase social isolation, decrease locals' chances of staying in the neighbourhood, undermine immigrant reception, and constrain housing choice (Bolt, 2009; Bolt et al., 2010; de Koning, 2015; Lees, 2008).

As a result, social mix has been called a mask for gentrification (Lees, 2008), or indeed a form of gentrification itself (Davidson, 2008), where gentrification is understood as the production of space for progressively more affluent users (Hackworth, 2002), causing the eventual displacement of all or most of the pre-existing low-income residents (Bridge et al., 2012). However, this reading of mix tends not to fully capture how "difference produced and spatialized in the context of colonialisms" (Roy, 2018) plays a central role in the discourse and practice and in the production of displaceability on which it depends. As a planning common sense, social mix is actually a *non*sense. It not only fails to do what it claims, it also claims to solve a problem that is not actually a problem in German cities (i.e., ethnic segregation) (Harlander, 2012; Münch, 2009). Here I am in conversation with and echoing the formulation of Sara Ahmed's (2004) characterization of anti-racism as non-performative: as she puts it, some declarations do not do what they say. I do not mean to suggest, however, that social mix is posited as performative: the power of it is ultimately in the doing rather than in the saying of "mix," though the latter enables the former. Rather, the taken-for-granted basis for the discourse and practice of social mix is fundamentally faulty, and thus exemplifies the kind of obvious contradiction that El-Tayeb (2016, p. 40) argues make the deep structures of racism visible.

As a result I approach social mix as a "racialized technique of recasting relations of domination," as Kipfer and Goonewardena (2014, p. 203) put it in their research about Toronto and Paris. The contradiction and the racialized displaceability that social mix requires is anchored in national law, through an exception to the General Act on Equal Treatment, which establishes that discrimination in housing is permissible when "neighbourhood balance" (i.e. mix or a lack thereof) is a concern (Bundesministerium der Justiz und für Verbraucherschutz, 2006; N. Ha, 2014b). Sybille Münch (2009), a public policy scholar in Germany, argues that the myth of mix is enabled by the taboo around conducting real research into discrimination in Germany. In the absence of discrimination research, social mix masquerades as a strategy against segregation, isolation, and social exclusion, though it deflects attention away from the actual causes of urban spatial inequality (Bolt et al., 2010; Lees, 2008) and obscures the needs of vulnerable people (Saville-Smith et al., 2015).

Conclusion

In this chapter, I have outlined the main theoretical and conceptual threads that weave throughout this book, in order to frame the chapters that follow. I began by naming my citational practice, through which I centre and aim to amplify the work of Black German scholars and German scholars of Colour that comprises the pathfinding critical race urban scholarship in and about Germany. In a context where some of the dominant modes of racialization and boundary-making around German identity involve the externalization of racialized "others" from Germany and from Europe, a long-standing taboo around race and racism has not only created a gap in research, it has contributed to the migrantization of many people who are at home in the country. I outlined Fatima El-Tayeb's definition and analysis of racialization in the German context, which I draw on extensively, and sketched out the importance of "migration background" and anti-Muslim racism as locally specific currents. I also touched upon integration as an influential impossibility in a context where racialized people are considered to be fundamentally out of place.

I then defined racial capitalism as an important concept and system in my analysis of development in Wilhelmsburg. I will pick up this thread again in Chapter 3, where I explore Wilhelmsburg's history of stigmatization through this lens. Drawing particularly on the work of geographer Laura Pulido, I have linked racial capitalism to two key concepts, environmental racism and white supremacy, which are, respectively, by-products of racial capitalism and forms of racism that produce devaluation. I argued that these systems and processes produce and are reproduced by urban space and identified several spatializations that are key to this book. The racialization of space, segregation, and the ghetto depend upon notions of control of racialized people through containment in urban space. In contrast, racialized displaceability underpins the attempted control of racialized people through dispersal and exclusion. To conclude this chapter, I focused in on the discourse and practice of social mix, as an urban planning common sense in Germany that depends upon racialized displaceability to function. In the following chapters, social mix will emerge as a crucial planning technique in Hamburg-Wilhelmsburg, but also as a conflicted basis for hope from the perspective of residents. The disciplining notion of integration is always just under the surface of social mix in this context – however, racialized Wilhelmsburgers question who needs to integrate with whom.

Before moving to what I found in Wilhelmsburg, I discuss how I came to be there in the first place. In the following chapter, I establish my study as an anti-racist intersectional public ethnography. I describe my research methods and the scope of the study, in which my own presence and reflexivity as a researcher was crucial. The project was not without challenges and tensions, created in part by the very context of racialization that I have established in this chapter. I explain how I handled the opportunities and challenges of the study before I dive into racialized residents' analyses of development and change in Hamburg-Wilhelmsburg.

2. Methodology and Methods

Introduction: A Canadian in Wilhelmsburg

I sat down for a beer in Stübenplatz with local newspaper reporter Hannes Lintschnig in July 2018, shortly before returning home to Toronto. Perched on concrete planters in the busy hub of transit, shopping, and hanging around, we smiled at the appropriateness of getting together where much of Wilhelmsburg meets, and chatted about my research. The article that Hannes proposed was not about my research, however. Instead he said: "What people here will really want to know is how does a Canadian end up doing her doctorate in Wilhelmsburg?!"[1]

How I came to this project was indeed the most-frequently asked question when I was out and about during my research. I would usually explain it as happenstance: I had long studied the German language, and in the middle of my master's studies I was part of a study tour of Hamburg with eleven other students from Canada. As soon as we arrived, we began to hear about Wilhelmsburg; the island's development and re-development was a hot topic for the city. Because I was studying education and had worked in immigrant-serving organizations, a short internship was organized for me at the Haus der Jugend Wilhelmsburg, a centre for kids and youth. The rest, as they say, is history.

Yet as I briefly mentioned in the Introduction, this project was not really the product of happenstance, or simply of my privileged mobility: it was one specific thing that I heard in Hamburg that eventually led me to this precise research. In a talk about Hamburg's urban development and renewal strategy and priorities, the then head of planning told our study group that in order to solve Wilhelmsburg's problems, the population needed to be transformed. I was taken aback by this framing of the 'problem,' and I wondered: "What precisely is this "population" of which he speaks? And *why* do they need to be "transformed"?"

I spent most of my time in 2011 in the Reiherstieg area, the northwest part of the island that is sometimes called Old Wilhelmsburg, observing and having long conversations with the generous staff and volunteers at the Haus der Jugend Wilhelms-

1 See Lintschnig (2018) for the article.

burg, which is a neighbourhood centre for children and youth. Wilhelmsburg-Reiherstieg was covered in beautiful old buildings in need of a facelift. A burnt-out car sat on a corner for the entire time I was there. Commuting by bike each day from the other side of the river where I was staying, I smelled the neighbourhood's distinct, acrid smells of industry and pollution, and I compared the island to downtown Hamburg, and to other parts of the city. When I returned home to Toronto, it was with a nascent critical analysis of how Wilhelmsburg was talked about by people in positions of state power and by the media.

The master's thesis I then conducted was a precursor to this study. Whereas that research focused on the notion of the "population" in the statement "the population must be transformed," my doctoral research focused on the notion of 'transformation' and on what these developments *do*, particularly for the racialized "population." In this chapter I will describe the project's research questions, methodology and methods.

Research questions and objectives

The research questions for this doctoral study were:

a) How do racialized[2] residents of Wilhelmsburg experience and engage with the recent interventions in their neighbourhood?
b) How do racialized people figure in local planners' and politicians' decision-making?

In practice, the first question receives more focus in this book than does the second. This is a result of the breadth and depth of findings on residents' experiences, which overflowed the limits of my questions as I initially understood them. In keeping with the anti-racist critique that frames the research, I have given this data centre stage, with the data on the second question playing an important supporting role.

To match the research process to the anti-racist critique embedded in the research questions, I planned a project based in anti-racist and intersectional public ethnography methodologies. In these methodologies, which I will describe in the following section, researcher reflexivity and positionality are considered to be crucial aspects of knowledge production. Researcher subjectivity is not to be denied or avoided; the researcher is called upon to reflect upon their presence in every aspect of the research, making their choices, challenges, and analysis explicit and transparent.

2 See the previous chapter for a detailed definition and discussion of how I understand racialization in general and in the German context.

This imperative to be reflexive and transparent is why I chose to open this chapter by narrating how I came to this research. From the outset, this has been outsider research. I am a straight, white cis-woman born and raised in Toronto, which is in the Dish with One Spoon Territory, on the traditional territory of the Anishinaabe, the Haudenosaunee, the Huron-Wendat and Petun First Nations, and most recently the Mississaugas of Credit River. My family are Anglo-European settlers on Indigenous land, and have moved within what is now called Canada, but we are many generations removed from the experience of international migration. I come from an upper middle-class background, the child of university-educated professionals. I am a native English speaker and I speak German well, mostly (I am told) without an accent. As will become clear in the following chapters, many of these aspects of my identity and experience position me as an outsider to Wilhelmsburg and to the community with whom I conducted this research. I chose a methodology and methods that would put this outsider status to use and not shy away from the issues attached to it, as I will discuss in more detail below. First, I will describe the methodological principles and assumptions embedded in an anti-racist, intersectional public ethnography.

Figure 5: The market at Stübenplatz, a hub of activity in Reiherstieg. Twice weekly, the market features an extensive selection of inexpensive clothing as well as fruit and vegetable stands, meat, cheese, and baked goods

(photo by author, 2017)

Methodology: An anti-racist, intersectional public ethnography

Anti-racism has been integral to my research motivation since my earlier work on IBA Hamburg. Anti-racism aims to "identify, challenge, and change the values, structures, and behaviours that perpetuate systemic racism and other forms of societal oppression" (Dei, 2005, p. 3). Anti-racism research methodology begins from the assumption that research is not neutral, but always has interests embedded in it, and seeks to integrate anti-racist principles into all aspects of the research process. According to George Dei's (2005, 2015) conceptualization, in practice this means troubling every aspect of research, beginning from what kinds of questions are to be posed, by whom and why, to who is allowed to speak and produce knowledge that is considered valid, authoritative, and empirical.

I see this as an anti-racist research project because, consistent with Dei's (2005, p. 11) definition, it centrally asks about racial domination and social oppression, "with an objective of providing local subjects with an opportunity to speak about their experiences within the broader contexts of structural and institutional forces of society." However, the first question I considered when planning the project was whether I should do it at all. There is a long history of exploitation and extraction of racialized communities in the name of knowledge production (L. T. Smith, 2012; Stewart-Harawira, 2013; Sudbury, 2007). Dominant research tends to pathologize and stigmatize communities that are already marginalized (Dei, 2005). Even critical research often fails to show communities in their full complexity (Pulido, 2008), and reproduces categories of difference in dominant terms (see for example Arslan, 2015 review of Hinze, 2013).

As a result, the principles of anti-racist research include challenging power relations, negotiating power, control and interpretive authority, transparency, accountability, relationality, and treating theorizing as something that happens in and with community (Dei, 2005). To this I add the compatible value of reciprocity, which Laura Pulido (2008) describes as a defining feature of activist scholarship. Intersectionality-informed research methodology adds an explicit focus on investigating and acting on the multi-dimensional nature of people's lived experiences, based on the a concept of intersectionality as developed by Kimberlé Crenshaw (1991). Crenshaw (1991, p. 1245) argues that peoples' lived experiences are wrapped up in "multiple grounds of identity" and numerous axes of oppression including sexuality, gender, ability, class, race, and immigration status. Attending to people's actual experiences of the social world along these multiple grounds increases the likelihood of comprehending oppression in a nuanced way, and thus being able to take appropriate action (Crenshaw, 1991).

Intersectionality-informed research assumes that people's interpretations and navigation of everyday life and power relations are shaped by multiple social categories, oppressions, and privileges. Authority on what problems and axes of identity are important lies with participants themselves (Hunting, n.d.). The researcher's role is to build intersectional analysis into each step of the research process, asking questions that allow people to discuss their experiences in their own terms (Bowleg, 2008). As Billies (2010) demonstrates, the researcher plays an active role, making an analytical and conceptual contribution to the relationship.

The first conceptual contribution that I make is in trying to formulate and realize the project in such a way that it resists the objectification of racialized people. To do this I emphasize two things: Firstly, racialized people are producers of *analysis*, not just "affectedness" (El-Tayeb, 2016, p. 22, translation by author). Fatima El-Tayeb (2016) argues that racialized people's perspectives – where they are valued at all in academia – tend to be invited exclusively to report upon lived experiences of a particular issue (affectedness), and not interpretation of the issue. This echoes Julia Chinyere Oparah's (formerly Sudbury, 2007) insight that knowledge industries produce two distinct kinds of subjects: one the authoritative expert (researcher) and the other the raw materials (participants, interviewees, etc.). Anti-racism in research requires seeing people as creating and living theory (Dei, 2005). As a result, in this study I approach Wilhelmsburgers as authoritative thinkers and analysts. Secondly, and following a similar logic, I highlight practices of resisting and speaking back to discourses and structures of domination (Dei, 2005). In this way I consider racialized people to be complex and diverse actors who have agency in their neighbourhood, city and beyond.

I also articulate the fundamental concerns of the project as being about the discourses and practices of planning and governance, and thus situate it as a form of domination studies (Dei, 2005), rather than a study of racialized people per se. An illustration of this is that I have often been introduced in academic settings during my work in Hamburg as, "Julie, she does research on migrant communities." When this would occur, I would say, "No, I do research about the effects of racism in urban planning and development on people who live in stigmatized neighbourhoods." I understand this distinction as an important one that resists slippage into the dominant realm of research about marginalized communities as *objects*. Certainly, the lives of racialized Wilhelmsburgers are central to this project, but as people who know and produce knowledge on a pressing research topic. My hope is that this distinction is clear, significant, productive, and of interest to multiple audiences.

I approached this research as a form of public ethnography. What makes an ethnography *public*, is that it has an explicit critique that aims to challenge social injustice and speak to an audience that is not solely academic (Bailey, 2013; Tedlock, 2005). There is anti-racist critique embedded in the research questions and focus of the project. My hope is that the findings will, however slowly, reach politicians and planners in Hamburg, Wilhelmsburg residents, as well as a scholarly audience and that, as Henderson (2016) suggests, the ethnographic methods used will help to make the research accessible to an urban planning audience.

Ethnography, according to Sarah Pink (2007, p. 22), is not a method of data collection as it has historically been understood, but rather "a process of creating and representing knowledge (about society, culture and individuals) that is based on ethnographers' own experiences." It is a reflexive methodology; the researcher does not claim to report objective truth through research, but rather aims through reflection on power, relationality, and representation to offer versions of their experiences that are as faithful as possible to "the context, negotiations and intersubjectivities through which the knowledge was produced" (Pink, 2007, p. 22). A researcher may use the "traditional" ethnographic methods of interviews and observation, producing textual and written data, but also objects, images, sensory and immaterial experiences are part of ethnographic knowledge production (Pink, 2007, 2008).

Ethnography according to this definition is highly compatible with the principles of anti-racist and intersectional research. It assumes that the researcher has a social location and a way of viewing the world that need to be reflected upon and illuminated in the research. In the following sections I describe in detail the methods used and discuss some of the methodological issues that arose in the process.

Methods

This research was conducted from 2016-2019, with 10 months in Hamburg-Wilhelmsburg and the rest in Toronto. In the summer of 2016, I conducted two weeks of exploratory research to gauge the feasibility of the project idea. I reconnected with people I knew in Wilhelmsburg from my time there in 2011, and reached out to a few prominent activist organizations. I met mainly white-dominated organizations at that time, so I did not shape my dissertation proposal based on their agendas, but rather took their questions about how I would undertake the research and their interests in learning more about the diversity of experiences in Wilhelmsburg and brought those together with the existing empirical and theoretical literature by Black German scholars and German scholars of Colour. This became the basis of my research proposal, which I treated as provisional until I returned to Wilhelmsburg the following year and concluded based on the interest I encountered from potential participants that the topic was indeed of some relevance.

The methods I used in this project included ethnographic interviews, participant sensing, photography, and archival research. Ethnographic interviews are a mainly oral method of data collection, in which the researcher listens and asks questions of someone to produce verbal descriptions and theories (Hammersley & Atkinson, 2007). Participant sensing is a multi-sensory expansion of the ethnographic method of participant observation, in which a researcher collects data by being present somewhere, taking part in events and activities, and reflecting on those experiences (Pink, 2009). Photography in this study was a researcher-driven form of reflecting visually on the materiality of Wilhelmsburg through the lens of my research questions (Pink, 2007). I initially planned to conduct walking interviews and ask participants to direct me to take photos; however, I quickly found that participants needed to meet on their lunch breaks or in between other responsibilities. In keeping with the anti-racist research principle of responsiveness to the needs and priorities of research participants, I let the idea go, and instead conducted more traditionally structured interviews and took photos myself throughout the research process.

Archival research

My archival work was conducted from 2016-2019 through in-person and online searches of materials of possible relevance to Wilhelmsburg's history and development. This included many visits to the city's public library at the University of Hamburg and to Wilhelmsburg's local organization for historical documentation and education, Geschichtswerkstatt Wilhelmsburg & Hafen. There I was helped extensively by Margret Markert, who dug up and sent me dozens of documents and referred me to books and films. I also made regular trips to the local bookstore in Wilhelmsburg, Lüdemann, which carries a large collection about the island and always features the newest publications.

At the same time, I carried out archival research online via a number of key sites, where I used systematic searches about Wilhelmsburg as well as targeted, subject-focused searches to identify relevant materials. Key online sources included the city of Hamburg repository for legislative documents, the Ministry of Urban Development and Housing website, the Hamburg.de history portal, and the IBA Hamburg website. I also searched the online archives of the *Hamburger Abendblatt, Tageszeitung, die Welt,* and *Der Spiegel* for media reports on Wilhelmsburg. Throughout the project I maintained a regular search of the *Hamburger Abendblatt*, and read the Wilhelmsburg local newspaper, the *Wilhelmsburger Insel Rundblick.*

Interviews with Wilhelmsburg residents

From May to August 2017, I conducted ethnographic interviews with 19 Wilhelmsburg residents who have lived and worked in Wilhelmsburg from 14 to over 50 years. Ethnographic interviews can take many forms; in this study they were generally solicited, loosely planned and therefore somewhat formal, and semi-structured. Eight of the participants identified as women, and eleven as men. They ranged in age from 24 to over 50 years old, and have a variety of paid and unpaid occupations: I interviewed one lawyer, one orthodontist, two people who work for Hamburg waste services, three full-time mothers, one person on disability, two students, one small business manager, one small business owner, one office worker, one retail worker, two social services workers, one employee of city transit services, two pastors, one elder-care worker, and a teacher.

Five participants were born in Hamburg to immigrant parents, and 14 immigrated to Germany either as children or as adults. Participants told me about family and personal connections to Turkey (14), Bulgaria (1), Ghana (1), Nigeria (1), Palestine (1), Spain (1), and Portugal (1). Two of the participants with roots in Turkey were Kurdish. Two participants self-identified as Christian, one as Alevi, and eight as Muslim; another eight participants did not disclose their religion, if any.

The interviews lasted between 45 minutes and two hours, with the majority around 75 minutes. Interviews took place wherever was convenient for the participant, mainly at their workplaces and in cafés. I conducted seventeen interviews in German, and two in English. I was prepared to work with an interpreter for other languages as necessary, but found that the inclusionary criteria of 10+ years in Wilhelmsburg meant that most potential participants spoke German. Indeed, when I first arrived, the Turkish-speaking Wilhelmsburgers who gave me advice on recruitment strategy rolled their eyes at the mention of interpreters. My assumption that one might be needed spoke to them of the stereotype of "failed integration" in Wilhelmsburg that had little to do with their daily reality. Based on their guidance I resolved to see how it went with German; I stopped at nineteen interviews simply because of the limits of my capacity as a solo researcher.

I recruited most participants through community organizations, followed by snowball sampling. My first two interviewees were found through spontaneous conversations, however, at a convenience store and at a printshop; I mentioned that I was a researcher visiting from Canada, and upon hearing the topic the shopkeeper and owner respectively said, "You can interview me!" There were two inclusionary criteria: 1) that they had lived on the island for more than 10 years (and had therefore lived through the recent large interventions), and 2) that they related in some way, however conflictual, with the concept of *Migrationshintergrund*. In practice, I stretched the first criteria to include one person who had recently moved to Wilhelmsburg, and one who had recently moved away. This stretch made sense because both had done neighbourhood-engaged work in Wilhelmsburg for decades, one working as a counsellor serving migrants and low-income people, the other growing up in and around their parents' local small business.

The inclusion criteria of relating to *Migrationshintergrund* was fraught with tensions because of the label's status as a racializing concept. A census category and common term in German public discourse, it labels certain people as distinct from the imagined normal mainstream, based on a family experience of migration but regardless of whether they themselves have immigrated (Institut für Migrations- und Rassismusforschung Hamburg, 2011). More often than not it is used simply to distinguish "others" from people considered to be white, and for German-born children or grandchildren of immigrants, it tends to evict them from their own country (see Sow, 2018 for a plain language explanation). I discussed *Migrationshintergrund* in detail in the previous chapter, as it is an important feature of racism in Germany, but it bears brief discussion here because it was a challenging aspect of recruitment. Sara Ahmed (2007, pp. 149–150), writing about whiteness, suggests that we might expect that "any project that aims to dismantle or challenge the categories that are made invisible through privilege is bound to participate in the object of its critique." I often had Ahmed's observation in mind during recruitment of participants. I used the word *Migrationshintergrund* in my outreach, while attempting to unsettle it either by referring to it as "so-called" or by putting air-quotes around it. My written outreach documents also included other terms that have been embraced by some communities, such as People of Colour, Black Germans, and migrants; these terms have a different political tone because they have been taken up as terms of resistance and self-definition, but are less common (see again Sow, 2018).

Where people self-identified as potential participants it was relatively unproblematic, as *Migrationshintergrund is* a common term and I knew that there would be space within the interviews to disrupt the label and talk about self-identity. It was more troubling when I asked people and organizations to suggest possible participants, in effect asking them to identify other people with this label. I reached out to organizations that explicitly identify as working with residents who are migrants, or who are racialized in the German context, such as the local Turkish parents' association and "integration" service providers. Others, such as churches and youth centres, work with residents of various backgrounds. A single contact also opened some doors for me: a retired local doctor and long-time activist in Wilhelmsburg introduced me to several people who I then interviewed and who suggested further contacts. It was often possible to draw attention to the problematic nature of *Migrationshintergrund* in conversation, and thus to disrupt the way the category worked in the moment. Yet the normalization of the concept, its reach, and the challenges presented when rejecting it are precisely symptomatic of the ubiquity of race and racialization in the German context.

I focused my energies on ensuring that a disruption of power relations and of dominant discourse was present in the interviews. The overall topic of the interviews meant that participants often narrated starkly non-hegemonic analyses of conditions in Wilhelmsburg from the outset. For example, participants would often say things like, "Wilhelmsburg isn't like they say it is," early in the interview, so that very quickly we were talking about discourse and power, and about a "they" versus a "we" with different perspectives, interests and knowledges. Critical conversations were thus often underway before I could even turn on my voice recorder. Oppression, inequity, and privilege were talked about quite directly in all interviews.

I aimed to interview residents who roughly represented the diversity that is collapsed within the category of *Migrationshintergrund* in Wilhelmsburg. I particularly had diversity in terms of birthplace, religion, gender, class, and sexuality in mind during recruitment. Based on the concept of intersectionality I anticipated that Wilhelmsburgers would be variously situated in relation to the changing neighbourhood, and indeed participants talked about religion, gender, and class-specific experiences that are quite striking and illuminating. Racialization does not operate independently, but rather in "complex and shifting interactions" with other systems of oppression (El-Tayeb, 2011, p. 125). It was important to reflect this understanding from the outset.

I found that Wilhelmsburgers in fact pushed me towards a participant sample that reflected some of the history and diversity of the island. People with roots in Turkey are the largest racialized group in Wilhelmsburg, but they are not the only one (Bezirkamt Hamburg-Mitte, 2015). Wilhelmsburgers often asked me who I had talked to so far and urged me not to forget particular communities such as "the Portuguese" and "West Africans." As I will discuss in the following chapters, I saw this as resisting the conflation of "migration background" with Turkish and Muslim and asserting an oppositional version of Wilhelmsburg, which has a diversity of which residents are proud. The demographic that I reached most quickly was Turkish-speaking and mainly men, so I began doing purposive outreach to diversify the sample, contacting, for example, women's organizations, churches attended largely by African and Black German residents, the Hamburg branch of the organization of Black people in Germany, and Hamburg-wide LGBTQ organizations.

While this study involved just a small sample of Wilhelmsburg residents, and thus cannot be considered representative in general, one of the particular limitations is the lack of participants who self-identified (to me) as LGBTQ. This is unfortunate because the literature on German cities demonstrates that racialized LGBTQ people are often treated as not existing in urban space (or at all) (El-Tayeb, 2012; Petzen, 2009), and that white homonationalism can play a significant role in the gentrification of racialized neighbourhoods (Haritaworn, 2015). I draw on this rich literature to address this gap where it arises in the following chapters. I found that the recruitment strategies I used in general did not work when it came to LGBTQ residents. I could not find any LGBTQ organizations in Wilhelmsburg, snowball sampling did not yield any results, and by the time I was in touch with a city-wide organization, my limited time in Hamburg proved a barrier to setting up a meeting. I am also not certain that I created the conditions in interviews wherein LGBTQ residents would have self-identified to me, as after the first few interviews I was reluctant to ask about aspects of identity that participants did not first raise themselves. At first, I was asking directly about how people self-identified in a number of ways, but I experienced the results as awkward and taking away from the conversation. Language was likely a factor, as I was not confident about how best to phrase these questions, but I also had a strong feeling that it was best to simply let people tell me what they wanted to about themselves. While letting participants tell me what axes of identity were important is consistent with intersectionality-informed research (Hunting, n.d.), upon reflection I think the line between my methodological choice and my own (straight, white, WASPy) sensitivities was a bit blurred, as gut feeling was a key driver of the choices I made within interviews. There are multiple factors that inform the choices that researchers make at any given time, related to questions and methodologies, personal history and interests (Pink, 2007), yet because I do not identify as LGBTQ myself, I wonder about the source of my gut feelings here. Next time, I would start outreach to city-wide organizations earlier, giving contacting, scheduling, and relationship-building more time.

Other challenges in this phase of the research included my language skills – when I listen to the recordings, I notice a real difference in my fluency between the first interview and the last, particularly in my phrasing of spontaneous questions in response to what participants said. The interviewees were very generous with me. With the exception of one person who did not want to be recorded, all interviews were audio-recorded for later transcription, and I also took notes. To thank people for their time I gave them a 10 € gift certificate to a nearby market, pharmacy, or café. Interestingly this seemed at first to offend participants, who then wanted to make it clear to me that it was not necessary or expected. I tried explaining where I was coming from, that I offered a small token because I was taught that it was respectful research practice. That seemed to be appreciated and interesting rather than offensive, so I continued doing it. People participated in interviews either because they liked what I was doing, or because someone they trusted introduced me, so I think I had inadvertently offended the relational basis of the encounter with the early participants. When I situated the token of thanks in relation to me as a researcher, the gift certificate seemed to take on a different, acceptable, meaning.

I discussed anonymity, confidentiality, and how the participants' words would be used at the beginning and the end of each interview. The majority of participants did not want to remain anonymous: of nineteen interviewees, just five preferred not to be named. Most of the names in the text are therefore participants' real first names, and I indicate where I have assigned a pseudonym at the participant's request.

Photography and participant sensing

I used the methods of participant observation and photography throughout the study. Alongside interviewing, participant observation is considered a "classic" ethnographic method, and is based on the premise that "being there" in a particular context can produce knowledge (Pink, 2009). The definition of participant observation that I worked with was in fact Sarah Pink's (2009) re-conceptualization of it as participant *sensing*, which is a multi-sensory and reflexive method of learning through embodied emplacement. In participant sensing, a researcher lives the unplanned experiences of everyday life in a particular place, and seeks out ways of sharing other people's experiences by purposefully joining in on activities and events. The expectation is neither of distanced, objective observation, nor of sharing the "authentic" experiences of others, but rather of having one's own engaged embodied and emplaced experiences in an environment that one wishes to know better (Pink, 2009).

I lived in Wilhelmsburg for four months in 2017 and six months in 2018, in Reiherstieg, which is one of the more densely populated parts of the island. I lived in a room facing northwards towards the port lands and city centre, with a view of the Spreehafen, Ernst-August Canal, and the trucks barreling along the road into the port. This was my base from which I had my daily life, as well as purposeful effort to participate in anything to do with planning and development in Wilhelmsburg. I attended, for example, three consultation events hosted by IBA Hamburg, which is now a municipal corporation responsible for project development. In 2017 and 2018 there were numerous meetings about the development of the "Spreehafenviertel," a large redevelopment across from my apartment, and I attended all those to experience the events and collect the maps and reports produced through the process. I also went on two IBA Hamburg tours: one focused on projects completed in 2013, led by an IBA-recommended guide, and one focused on works in progress, led by IBA staff during the 2017 national conference on urban development policy. These opportunities to experience IBA Hamburg's perspective on development in action were so rich that I did not find it necessary to also interview a representative. I also participated in a community forum about the forest slated for development into the "Spreehafenviertel," a church service where people spoke about living conditions in Wilhelmsburg, an open meeting held by elected politicians on the topic of development, and countless other events, celebrations, concerts, bike tours, organizing meetings, and more. I took notes about these experiences either during or after.

In between these activities and the work of interviewing, daily life unfolded. I shopped at the market on Wednesdays and Saturdays, went to local cafes, sat outside in the good weather and hid away in the rain. I joined a local gym, and a local choir, with self-care in mind, and found that it enriched my life and my research. I learned an enormous amount through this embodied everyday life in Wilhelmsburg, as will come up frequently in the coming pages. A reflexive awareness of my sensory experiences allowed me to note and eventually incorporate sounds, odours, feelings, and sights into my analysis.

While living in Wilhelmsburg I carried a camera with me virtually at all times. The camera on my mobile phone produces good quality images, so long as they are not enlarged much. Using it had the advantage of drawing little attention to myself, because of the ubiquity of mobile phones and cameras. At times I also carried a digital SLR camera, particularly to photograph places that residents told me about, and on regular bike rides around the island. The digital SLR produces much higher resolution images that then allow for enlargement, but it is also a much larger camera that draws attention to me as a photographer. This was not necessarily a bad thing, as I was not trying to photograph covertly; it was more an issue of my own comfort with feeling highly visible at any given time.

I approached photography within this research as a practice of paying attention to and making visual note of the materiality of Wilhelmsburg. The gaze was of course my own (Pink, 2007), rooted in my research questions, which I loosened to think in terms of key words such as change, development, racialized people, impact, and planning in Wilhelmsburg. I photographed what caught my eye, as well as what I knew to be physical artifacts of the processes I wished to question, such as IBA Hamburg's developments, new student residences, new cafés, and a disappeared refugee camp.

Interviews with residents shifted how I thought of the role of photography in the project, and introduced a collaborative component. I began taking photos of the specific places that participants told me about in interviews. This often took me to places that were new to me, or that I had visited and not photographed, bringing together my gaze with participants' knowledge and experiences. I was also so often told in interviews that Wilhelmsburg had been misunderstood and mischaracterized that I began to ask myself each time I lifted my camera: "What version of Wilhelmsburg will this image reflect?" Participants were anxious that I should communicate that Wilhelmsburg is more than its "bad image" and stigmatization, and I wondered how to reflect complexity and critical analysis through my photography.

One of the challenges was that "race" operates as a visual technology. As Jennifer A. González (2008) discusses in her work on the disruption of "race" in installation art, "race" produces "economies of visibility" (p. 5) in which race is marked on bodies – and, as Sherene Razack (2002) and others also note, *spaces*. Given this visual character, and given that images have no fixed meanings, but simply allow for interpretation of what is visible (Pink, 2007), a major ethical risk seemed to be producing images that would serve the dominant discourses about Wilhelmsburg. Further, visual ethnographic methods have often been used to objectify racialized and colonized people, including in the name of producing knowledge and "evidence" (Pink, 2007). My own gaze as a white, outsider researcher was part of this challenge, as photographers produce images not only out of their explicit intentions, but out of their most basic ways of thinking and perceiving, which are socially produced, culturally ingrained, and imbued with the structures of power (Pink, 2007, drawing on Bourdieu).

Yet because racial discourse often depends on visual discourse, visual methods can be a useful tool for disruption and revision (González, 2008). The strategy that I decided to adopt, alongside the methodological demands for reflexivity, transparency, accountability and so forth, was being selective in what and where I photographed and situating the images within (my and participants') critique and analysis. I took no portraits of people without their consent. As I was generally interested in spaces, buildings, and infrastructure to complement the individual and specific narratives of interviewees, this was not terribly difficult; it was in public spaces, in contexts where there would be an expectation of being seen and observed, that my photos at times included people. I also focused on the direction of resident participants, as I describe above. Finally, I contextualized and continue to contextualize the images when I display them by accompanying them with text. Pink (2007) notes that this dampens the viewer's creativity and the range of possible interpretations they might have. While this would be considered negative in some contexts, here it is helpful; I wish to guide the interpretation of images in this book just as I guide the interpretation of participants' words and my own sensory experiences.

A vivid conversation with a research participant illustrates how these strategies work together. To prepare the layout of posters for a closing event, described in the following section, Özden and I sat together looking at quotes from racialized Wilhelmsburg residents that I had grouped under the heading "the change isn't happening for the people who live here."[3] Özden, who is a graphic designer and owns a printing business, opened and enlarged a photo taken from my apartment window, of four men in a park on a sunny day. Three of the men appeared to be sleeping, and one looked towards a canal with the city's skyline in the background. I had taken the photo because I thought it illustrated something I frequently heard about: Eastern Europeans, particularly Bulgarians, who faced intense exploitation and housing precarity in Wilhelmsburg, and who lived largely outside during the summer, in parks like the one on my doorstep. After looking at the photo for a moment, Özden burst out laughing. I was taken aback, and for a moment just watched as he enlarged the image so that it formed the backdrop for the quotes. "It's perfect!" he said, "Look at them! This is exactly how it is!" I was surprised by the laughter and said so, and he told me: "It's perfect – the image, the title, what people say in the text here, it all goes together. I love this, I love this poster: the message cannot get lost!"[4]

The purpose of using photography and participant-sensing methods for this project was that they should complement and fit together with the interview methods I also used. In the final section, I discuss the second set of interviews, with

3 My translation of a participant's words.

4 See Figure 6.

planning professionals and politicians, and describe the community event I hosted at the end of the data collection phase.

Interviews with politicians and planning professionals

The final phase of the research in Wilhelmsburg was conducted from February to August 2018 and comprised of reporting back to resident interviewees, interviews with politicians and planning professionals, and a closing event to present my preliminary findings. In the winter I prepared a two-page summary about interviews with residents. I anticipated holding a focus group in which we would discuss and collectively analyze text from the previous year's interviews. However, though many participants responded to the summary, no one expressed a desire to meet and add to, correct, or elaborate on it. I had imagined the focus group playing a key role in the anti-racist methodological imperative to disrupt dominant researcher/researched relations, and in the location of interpretive authority within the community, so this felt like a methodological misstep. However, participants contributed their time, energy, and analysis to the project when they participated in interviews, and flexibility is also an important principle of anti-racist research. Further, my analysis is the greater part of what I contribute to the reciprocal relationship with participants, considering that the main thing people asked me for during the course of the project was to report back about what I learned and to have a community event at the end. The sharing of the summary document turned out to be an interim step in that report-back.

From March through June 2018 I conducted eight interviews with politicians and people involved in the planning process in a professional capacity. These included two elected representatives for Wilhelmsburg in the governing party at the national and city-state levels: Metin Hakverdi (*Sozialdemokratische Partei Deutschlands*[5] [hereafter SPD], member of the German Bundestag for Hamburg-Bergedorf-Harburg) and Michael Weinreich (SPD, member of Hamburg Parliament for Billstedt-Wilhelmsburg-Finkenwerder). Both were born and raised in Wilhelmsburg and continue to live there. It further included a representative of Wilhelmsburg at the district level, Bayram Inan (*Bündis 90/Die Grünen*[6] [hereafter *Die Grünen*] member of the district assembly for Hamburg-Mitte), who has lived in Wilhelmsburg since he came from Turkey in 1973 as a so-called guest worker. Chairperson of the Wilhelmsburg Council for Local Development, Lutz Cassel, has lived in Wilhelmsburg since roughly 2006. I also spoke with Heike Sudmann (*Die Linke*[7], member of Hamburg Parliament, responsible for urban development,

5 Social Democratic Party of Germany

6 Alliance 90/The Greens

7 The Left

housing and transit), who often speaks for and about the island because of her portfolio. At the ministry of urban development and environment I spoke with Michael Rink (Office of Regional Planning and Urban Development, Project Group Leap Across the Elbe), and informally with a colleague of his who was born and raised on Veddel, which is directly next to Wilhelmsburg, and who preferred to remain anonymous. Finally, I interviewed Sören Schaefer from *Perspektiven!*, which is a project of the Wilhelmsburg Bürgerhaus that organizes all of the development-related consultations required by local law. Schaefer is also a resident of the island.

In this phase I engaged in targeted recruitment, contacting specific people who I had already identified as potential participants. It was largely clear from the time I had already spent in Wilhelmsburg who the elected representatives were, and what bodies were responsible for development, and I therefore had a list of people with whom I wanted to speak. However, Hamburg has a multi-tiered and proportional system of representative government, which means that there are quite a few people elected by Wilhelmsburgers to sit on various assemblies and committees. As time did not permit me to interview all of them, I asked participants and other relations in the neighbourhood who I should talk to. The consensus was to prioritize those in the governing parties (the SPD and *Die Grünen* in Hamburg). Speaking to politicians on the more conservative end of the spectrum would have also been interesting, but given my focus on current planning and decision-making and my limited resources it was not a top priority to seek out those not currently making policy. There is also plenty to be gleaned from attention to the discourse and policy at the national level, where the ideologies of the Christian Democratic Union of Germany (CDU), Christian Social Union in Bavaria (CSU) and the Alternative for Germany (AfD) are highly visible and influential.

The format of the eight interviews was semi-structured, similar to my interviews with residents. However, between the oratory confidence of politicians and greater time pressure (most had less than an hour to speak to me) I found that asking effective spontaneous follow-up questions was more difficult. The first interviews strayed hardly at all from the dominant, economic narrative about the problems and solutions in Wilhelmsburg, skirting around the role of racism in past planning. To equip myself to get in more critical questions, I went back to my anti-racist, intersectional public ethnography methodology and prepared questions based on what I had heard from residents. Interviews took place at cafés and at participants' offices; I met Mr. Weinreich at home as he was recovering from knee surgery. I audio recorded all but two interviews and took notes for the others. These participants were not compensated for their time as they were speaking in their professional capacities.

Community event at close of data collection

Several research participants requested that I host a community event before I had to go back to Toronto, so that there would be an opportunity to discuss publicly and hear what other people had said. I planned the event for Saturday, June 30, 2018, shortly before the start of summer vacation, and held it in the event room of Don Matteo, which is a local family-run restaurant with several decades of history in the neighbourhood. All research participants were invited, as well as about a dozen Wilhelmsburgers who had supported and shown interest in the research in various ways. Fifteen people attended, mostly long-time residents engaged in community organizations and activism. Two were research participants; while several other research participants planned to attend, family responsibilities came up and took precedence. I titled the event *Mixed Feelings: Perspectives on living alongside and with each other in Wilhelmsburg.*[8]

The afternoon was a very fruitful discussion based on key themes from my interviews with racialized residents and informed by other experiences and interviews in Wilhelmsburg. With Özden's help, I had made eight posters that presented what I saw as key points of analysis, using quotes from interviews with residents and photos I had taken around the island (see Figure 6). People had a chance to look at the posters and then I made a short presentation about the theoretical basis for the project and what I thought I would emphasize in my dissertation. That was followed by two hours of rich and lively discussion, which I moderated while taking notes and fielding questions.

The discussion provided an opportunity for me to clarify my ideas, and participants also raised points that have made their way into this book. In particular, I have the attendees to thank for drawing my attention to the *long* history of Wilhelmsburg as an immigrant neighbourhood as it illuminates the ever-shifting terrain of racialization; the question of who defines and sets the terms and goals of "transformation;" and the problem of a lack of political will to fundamentally challenge the conditions of exploitation and deprivation faced by many of the newest migrants to the neighbourhood (in particular, "the Bulgarians," a category that attendees also productively unsettled, and that I will discuss in detail elsewhere). Attendees also expressed concern that the in-movement of more people who do not face structural oppression in terms of race, class, or immigration status might negatively affect practices of solidarity for which Wilhelmsburg has a reputation.

The event was an outlet for dissemination of my preliminary findings to a broadly sympathetic local audience, and the findings were overwhelmingly well

---

8 In German: Gemischte Gefühle: Meinungen zum Leben neben- und miteinander in Wilhelmsburg. Bericht zu einer Doktorarbeit

received. This was interesting in and of itself, as attendees primarily represented the dominant, white organizations on the island, and the posters and my presentation amplified voices that tend to be marginalized and lack representation in those organizations. As a result, while the event was motivated by accountability to racialized participants in my research, it also played a role in living up to the demands of my methodology by reaching a particular non-academic audience and using my privileged position to amplify local anti-racist analysis.

Conclusion

This book reports on an anti-racist, intersectional public ethnography that explores racialized residents' experiences and analyses of recent developments in Hamburg-Wilhelmsburg, and how racialized people figure in local planning. In this chapter I have described how a disturbing statement about transforming Wilhelmsburg's population in order to solve its problems brought me to this research. Informed by my master's thesis about how the IBA Hamburg project produced Wilhelmsburg as racialized and problematic, I aim in this book to address serious issues and gaps in how the neighbourhood's "transformation" is understood and evaluated.

I have positioned this study as the production of knowledge through my embodied experience as an outsider researcher (ethnography), and as a study with an explicit critique and the intention to reach multiple audiences (public). Anti-racist and intersectional methodologies inform the project's structure and goals, integrating anti-racist principles into all aspects of the project and approaching peoples' lived experiences as wrapped up in multiple grounds of identity and oppression. In this chapter I have described how my methods and research practices, particularly of accountability, reciprocity, and disrupting relations of power and authority within the research itself, reflect the demands of these methodologies.

I have also outlined the specific methods that I used: ethnographic interviews, participant sensing, and photography, complemented by an interim report to participants and a community event that used visual and textual tools to prompt discussion. I have described who I spoke to and why, and reflected on some of the possibilities and challenges that my methods presented. In particular, I reflected on the risk of reproducing what I wished to challenge through the use of the concept of *Migrationshintergrund* and the use of photography to communicate about the materiality of a space that has been racialized and stigmatized.

In the following chapter I analyze the production of Wilhelmsburg as a racialized and devalued space within the context of racial capitalism. Until recently, Wilhelmsburg had a distinctly negative image; it was labelled "the Bronx of the North," "the Balkans of the North," a "problem neighbourhood," and a "neighbourhood in crisis." The dominant narrative is that Wilhelmsburg's status emerged from an unfortunate but basically natural process of disinvestment by the city after a deadly flood in 1962. After the flood, a period of planning indecision ensued, followed by several decades in which Wilhelmsburg was treated simply as the city's "backyard." Residents argue, however, that the island's "bad reputation," as they often called it in interviews, hinged on the kind of people who were thought to live there: racialized people, low-income people and immigrants. In the next chapter I follow residents' insights and draw from the full breadth of data from this study, including government documents, secondary historical sources, media archives, participant sensing and photography in the neighbourhood, as well as interviews with residents, planners and politicians. Bringing these sources together with recent work by Black and People of Colour activists and scholars in Hamburg that illuminates the ongoing importance of colonialism to the city's accumulation of wealth through the differential valuation of people and land, I explore how Wilhelmsburg has been constructed in discourse and practice as an "other," non-European space, and thus as less valuable than other parts of the city.

Figure 6: "The change isn't happening for the people who live here:" One of eight posters displayed at the community event in 2018, with participant quotes and my photos

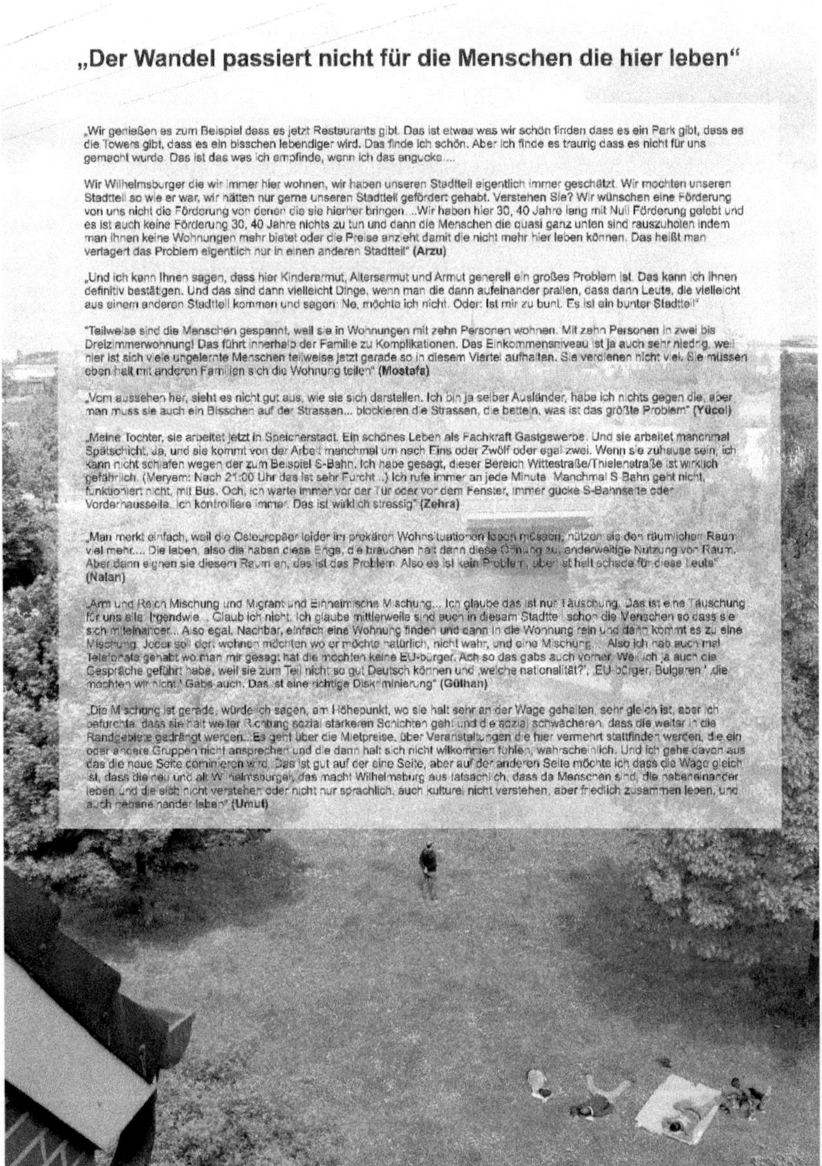

3. Producing the "Problem Neighbourhood"
Development of Hamburg-Wilhelmsburg in Racial Capitalism

Introduction

Hamburg-Wilhelmsburg is an expansive and complex place. The 35 square kilometre island is the largest of the Elbe islands, the largest river island in Europe, and home to 53,064 people (Statistisches Amt für Hamburg und Schleswig-Holstein, 2020). The island has been called "a patchwork of opposites" (Keesenberg, 1989, translation by author): it has agriculture, industry, and residential neighbourhoods in equal measure (Eckardt, 2017). The southeast of the island is dominated by bucolic farmers' fields and old farmhouses, punctuated by a cluster of high-rise apartment buildings (Figure 8). At the centre of Wilhelmsburg you find the mall, more high rises, single-family homes, and allotment gardens. To the northwest is the worker's row housing of Reiherstieg, or Old Wilhelmsburg, and beyond it lie container yards, heavy industry, and the port lands (Figure 9).[1] The landscape and built forms of Wilhelmsburg are tremendously varied; the island is loud and dirty, green and peaceful, nowhere is it just one thing.

The island's population is also tremendously heterogeneous (Hohenstatt & Rinn, 2013). In Wilhelmsburg, people with different languages, passports, religions, and to some extent socio-economic statuses, all live alongside each other with what residents characterize as a great deal of success. As Umut, who grew up around his parents' business in Old Wilhelmsburg put it: "It's what really makes Wilhelmsburg, that there are people who live side by side and who do not understand each other – not only linguistically, even culturally – but live together peacefully, and also live side by side" (Umut, interview, 2017).

The statistics about Wilhelmsburg reflect that many racialized people live here and many immigrants, as well as significant proportion who are counted simply

1 For a map of the locations mentioned here and throughout the book, see https://juliechamberlain.org/map-of-wilhelmsburg

as German citizens. In the most recent statistics, 31.8% of Wilhelmsburgers are listed as "foreigners" (with a passport other than German), and 60.8% are counted as people with "migration backgrounds." For youth under 18 years of age, the percentage with migration backgrounds is 78.3% (Statistisches Amt für Hamburg und Schleswig-Holstein, 2020). Wilhelmsburg is also the location of a cluster of homes in Georgswerder that is the city's only officially settled Sinti community (Journalisten der Henri-Nannen-Schule, 2015; Knödler, 2021). Income in Wilhelmsburg is on the lower end of the city average – not one of the lowest-income neighbourhoods, but just one step above (Statistisches Amt für Hamburg und Schleswig-Holstein, 2017). Wilhelmsburgers also live in a range of housing forms: single-family homes, apartments, subsidized housing; the mix is more varied than in other parts of the city (Statistisches Amt für Hamburg und Schleswig-Holstein, 2020, pp. 27–28).

Yet until recently, Wilhelmsburg had a very simplified image, and it was a negative one. It had been labelled "the Bronx of the North," "the Balkans of the North," a "problem neighbourhood," a "neighbourhood in crisis," and a "social hotspot" (cf. Adanalı, 2013; Barth, 2000; Brinkbäumer, 2000; Hamburger Abendblatt, 2000; Hohenstatt & Rinn, 2013; Twickel, 2011).

For decades, media headlines and government policy documents reflected perceptions of the island as a problem and reproduced its stigmatization. Often people from other parts of Hamburg had never set foot in Wilhelmsburg, despite the island's location a mere two S-Bahn stops away from the central station. Hamburgers "jumped three steps backward in horror" (Kopf, 2012, p. 102, translation by author) when they met someone who actually lived in Wilhelmsburg. It had a bad reputation: "a really bad reputation" (Casim, interview, 2017).

This bad reputation, over time, has led to Wilhelmsburg being the subject of strategic and targeted urban-planning strategies on the part of Hamburg city-state, particularly in the past fifteen years. The dominant narrative is that Wilhelmsburg's status as a "problem area" emerged from an unfortunate but basically natural process of disinvestment by the city after a deadly flood in 1962. After the flood, a period of planning indecision ensued, followed by several decades in which Wilhelmsburg was treated simply as the city's "backyard." Garbage was dumped there, dirty industry was dumped there, a high-rise complex was built there, and the population of immigrants and "socially weak" residents rose as Wilhelmsburg became one of the cheaper places to live in the city. Eventually, the island reached a tipping point, and the city was pushed by resident activism and national media attention to address the conditions that it had produced in Wilhelmsburg. As I will discuss in the following chapters, it does this today through strategies of "social mixing" that aim to re-balance the population by attracting the (white) German middle class.

Wilhelmsburg's "bad reputation" hinges fundamentally on the kind of people who are imagined to live there – immigrants, low-income people, the unemployed, people on social assistance – in short, people who are stigmatized and devalued in Hamburg and in German society. The residents I interviewed argued that the reputation is important to understanding current events, and traced it to a combination of factors that included actual conditions in which people of low socio-economic status were deliberately concentrated on the island, and racist discourse about the presence of migrants. This perspective does not contradict the dominant narrative, but rather connects the dots within it. It draws attention to the many steps in the process of devaluation of Wilhelmsburg, illuminating the assumptions about the value of who lives there that are required to treat Wilhelmsburg the way it has been treated by the city-state over the past decades. Crucially, the development of the bad reputation initiated long before the flood; the process of devaluation of the island dates back to the beginnings of industrialization, with the long history of Wilhelmsburg as an immigrant neighbourhood.

Following residents' insights, my objective in this chapter is to present a history of Wilhelmsburg that contextualizes recent developments and residents' experiences with them, but that also complicates the dominant narrative. To do this I draw from the full breadth of data from this study, including government documents, secondary historical sources, participant sensing and photography in the neighbourhood, as well as interviews with residents, planners and politicians. First, I present the dominant narrative in more detail, which offers one version of how and why Wilhelmsburg has historically been treated as a space of work and for waste. I focus on the details of this narrative, especially that the state continued to settle immigrants on the island *after* it was declared unfit for habitation, to argue that conditions of environmental racism were deliberately produced. Second, I draw on interviews with racialized residents to discuss the content of Wilhelmsburg's long-standing "bad reputation," and thus its devaluation in discourse and practice. Residents describe how the labels and images attached to Wilhelmsburg turn on, on the one hand, conditions of marginalization that are spatialized by the state (and, I argue, inherent to racial capitalism). On the other hand, racist discourses about migrant danger and inflated representations of crime contribute to views of Wilhelmsburg as not-quite-Germany, a pattern which I analyze in relation to the above-reviewed theories of the racialization of space, ghetto discourse, and anti-Muslim racism.

In section three I present a history of Hamburg in which Wilhelmsburg's bad reputation can be understood. Hamburg is a city that's wealth has been accumulated through the differential valuation of people and land in colonialism and in ongoing racial capitalism. Black and People of Colour activists and scholars in Hamburg have recently succeeded in putting the city's colonial legacy on the public agenda; I follow their insights to note how Hamburg's relationship to its "others," including "other" spaces like Wilhelmsburg, continues to be shaped by assumptions about "some people being worth more than others" (Adjei, in Della et al.,

2018, translation by author). Finally, I trace the devaluation of Wilhelmsburg further back in history, to racialized labour migration that stretches back to the 19th century and to the earliest choices by politicians and planners to dedicate the island as a space of work and waste. For over 150 years, racialized people have been recruited to come to Wilhelmsburg to do devalued work under inequitable conditions. This is, I argue, part of the normal functioning of racial capitalism. As I noted in Chapter 1, Cedric Robinson originated the concept of racial capitalism to capture the role that race and racism play in the production of value. I revisit this concept, drawing particularly on Laura Pulido's elaborations and applications of it, to argue that racial capitalism is a foundational structure on which Wilhelmsburg has been developed to date.[2] Wilhelmsburg's bad reputation, and thus the context for current developments there, is based in this devaluation of the people who live there. The following chapter, in contrast, turns to residents' own valuations of the island as a place of beauty, warmth and connection.

Figure 7: The view from my window facing north from Reiherstieg towards the Spreehafen and the city shows some of Wilhelmsburg's heterogeneity. The frame is full of the lush green of spring, and sheep graze on the dike, performing their crucial maintenance. In the middle of the image is one of the many transport trucks that fly by during the day on the way to and from the port; the dust and noise make it uncomfortable to keep the windows open

(photo by author, 2017)

2 Parts of this chapter appear in Noa Ha and Giovanni Picker's (2022) edited volume, *European cities: Modernity, race & colonialism* (Manchester University Press), where I situate this devaluation in relation notions of the 'European city' (Chamberlain, 2022).

Figure 8: The high rises of Kirchdorf-Süd seen from across the horse pasture to the south

(photo by author, 2017)

Figure 9: Workers row housing of Reiherstieg, recently renovated through IBA Hamburg, with container yard, industry and coal power plant in the background

(photo by author, 2017)

How Wilhelmsburg became a "problem"

> "Here on the former peasant islands of the marsh, Hamburg built its harbour, the economic heart of the metropolis. In immediate proximity to the sheds and quays, the shipyards and refineries, the steel and smelting, the storage, combing and roasting, working-class living space emerged"
> *(Zukunftskonferenz Wilhelmsburg, 2002, p. 5, translation by author).*

The dominant narrative about where Wilhelmsburg's negative image comes from, and thus why and how it requires planning intervention, comprises two main themes: the island as a space for work, and the island as a space for waste. The first part is the assertion that the city of Hamburg and its planners had not historically thought of Wilhelmsburg as somewhere people lived. Hamburg has been strongly influenced by Fritz Schumacher, who was an architect and the Head of Urban Planning for the city from 1909 to 1933. In 1920, not long before Wilhelmsburg became administratively part of Hamburg, Schumacher famously said that "Geest land is for living, and marshland is for working" (quoted in Zukunftskonferenz Wilhelmsburg, 2002, p. 5, translation by author). With this declaration he argued, in effect, that residential development in Hamburg should concentrate on the high ground to the north of the river Elbe, while the marshland that included the Elbe islands was best suited to labour; that is to industrial and harbour-related land uses. This framework is often referenced to explain the consolidation of Wilhelmsburg's image over time (cf. Behörde für Stadtentwicklung und Umwelt, 2005; Schultz & Sieweke, 2008; Zukunftskonferenz Wilhelmsburg, 2002); Schumacher's thinking remained influential long past the end of his tenure.

The second piece of the dominant narrative is that the deadly flood of 1962 was a turning point beyond which the city treated Wilhelmsburg as nothing more than a backyard dump (cf. IBA Hamburg, 2011b; Schultz & Sieweke, 2008). In February 1962, a storm surge from the North Sea broke through dikes at several points in northern Germany, including Wilhelmsburg. There was serious flooding in the middle of the night and 300 people were killed, 200 of them in Wilhelmsburg. Most of the Wilhelmsburgers who died were living in makeshift housing in a low-lying area, in fact, just in front of where I lived in Reiherstieg in a spot that is now slated for redevelopment. Much of Wilhelmsburg had been destroyed by bombing during World War Two, and those who lost their lives were largely those still without permanent housing 17 years later (Paech, 2008). The 1962 flood was a major emergency in Hamburg and beyond, and famously cemented the reputation of Helmut Schmidt, then Hamburg's interior minister, who would eventually become Chancellor of Germany. The emergency was marked by Schmidt's quick action to rescue and re-house people, which violated the German constitution at the time by deploying military personnel within German borders (Paech, 2008). The flood is a trauma that remains within living memory and is memorialized in permanent and temporary forms. Figure 10 shows a teardrop that fell from where it was hung in the forest in 2018, by activists who are protesting the redevelopment of the forest into the 'Spreehafenviertel' (see IBA Hamburg, n.d.-c; Waldretter, 2019). The teardrops reproduce the forest as a place of mourning, which has lain fallow since the event in 1962. It is still a place where vulnerable people live, as I will discuss in the following chapters.

The flood is frequently depicted by media and planning publications as the beginning of a downward spiral for Wilhelmsburg (cf. Behörde für Stadtentwicklung und Umwelt, 2005; IBA Hamburg, 2011b; Rowe, 2015; Schultz & Sieweke, 2008). After the flood, Hamburg moved to abandon housing in the northern part of the island, and was stopped only by extensive and successful protest by residents who wanted to stay (IBA Hamburg, 2011b). This remains an important part of Wilhelmsburg's activist history and self-image for some residents (cf. Zukunft Elbinsel Wilhelmsburg, 2012; Zukunftskonferenz Wilhelmsburg, 2002). Parts of the island were nonetheless declared unfit for anything but industrial and manufacturing uses, and the city later called an official halt to all renovation and investment in housing (Paech, 2008).

After the flood, as many as 10,000 people did move away, or as one planner puts it: "everybody who could afford to do so – only the low-income people stayed" (Kai Dietrich, quoted in Rowe, 2015, p. 40). Those who stayed experienced an acceleration in the siting of "undesirable elements from the city of Hamburg" (Schultz & Sieweke, 2008, p. 139), that over time included space-eating logistic centres and container yards, expanded highways cutting through the landscape, a sewage treatment plant and a toxic waste dump (Hellweg, 2010; Humburg & Rothschuh, 2018). What resulted has been called a "spatial and social cacophony" that gave outsiders an impression of Wilhelmsburg as a dirty, noisy, and unpleasant place (Schultz & Sieweke, 2008).

What followed the flood was the deliberate production of environmental racism in Wilhelmsburg; that is, the siting of undesirable city functions alongside people who were also considered undesirable (Gosine & Teelucksingh, 2008). Though the process of abandonment by the city tends to be depicted as unfortunate but natural under the circumstances, when the city of Hamburg renewed its planning of Wilhelmsburg as a space for work and waste, it continued to allocate housing to immigrants who were arriving to the city (Paech, 2008), thereby demonstrating that they were not valued the same as other Hamburg residents. From the 1970s it

Figure 10: At the top of Veringstrasse, a tear drop from the annual flood memorial has fallen onto the sidewalk

(photo by author, 2018)

Figure 11: A more permanent reminder on the side of a Reiherstieg building reads "flood 1962" and shows the water level

(photo by author, 2011)

also concentrated social housing on Wilhelmsburg (Eckardt, 2017). Together these fit even the narrowest definition of environmental racism and inequity (Bullard, 1996; Gosine & Teelucksingh, 2008).[3]

As local researcher Zeynep Adanalı argues (2013, p. 123), "the city itself created the 'social hotspot' of Wilhelmsburg." "Social hotspot" (*sozialer Brennpunkt*) is one of the many stigmatizing labels that the island has accumulated in the past decades (problem neighbourhood, Bronx), which refer above all to the kind of people who live in the neighbourhood. Residents themselves at times used the term "social hotspot," and I asked Arzu, who was born and raised in Wilhelmsburg, what it meant:

> Arzu: Wilhelmsburg is a social hotspot. And it has been for the last 50 years or so. But this part of the city was also given zero support ...
>
> Julie: What does 'social hotspot' mean, to your mind?

3 Though environmental inequity is under-researched in Hamburg, one study concluded that it is indeed present in the city, and that the percentage of "foreigners" in a neighbourhood (i.e. people without a German passport) was the greatest predictor of proximity to toxic emissions (Raddatz & Mennis, 2013).

> Arzu: The underclass was settled here – knowingly, I think – so that people with fewer prospects simply accumulated, lived here. They call that a 'social hotspot' because then of course many people don't work, or become criminal, or things like that. As it became prospect-less, as they stuck more and more of the underclass here, the neighbourhood became known for aggression, for criminality, and as the number of foreigners rose, and the underclass – I don't just mean people who don't earn much money, but also really the unemployed, the alcoholics, the drug addicted – so everything that sort of didn't have much prospect – was placed here. ... That was done, in my opinion, knowingly, controlled that way for years, decades, and now they're trying to shake it all up with this gentrification (Interview, 2017).

The creation of Wilhelmsburg as a social hotspot, a space where the undervalued in society are concentrated, is thus *intentional* in Arzu's view (it was done "knowingly"), *structural* (it is shaped by socio-economic hierarchy) and *systematic* ("it was controlled that way for years"). Wilhelmsburg's negative image flows from this, from the material conditions that are deliberately produced by the city-state. This, I will demonstrate in the following section, is the nature of racial capitalism: it depends upon such devaluation to create power and profit (Pulido, 2017). Arzu's analysis agrees with Adanalı's (2013): the city produced a situation that it now aims to "shake up;" not, however, by addressing the actual needs of the Wilhelmsburgers who, she indicates, are marginalized within a socio-economic order that is ableist, classist, and oriented towards paid employment as the measure of one's worth. Arzu instead argues that the changes that are happening in Wilhelmsburg are not for the benefit of those who already live there, as I will discuss in more detail in Chapter 6.

Analysis like this about the position of many Wilhelmsburgers within the prevailing social and economic order points to the fact that Wilhelmsburg has real issues that residents would like to see addressed. In the past, the island has also had some widely publicized issues that continue to affect how it is viewed from the outside. Yücel and Mohammed, who are both men who grew up in the neighbourhood (Yücel from the time he was 10), both told me in particular about a period in the 1980s when gangs developed that became quite notorious.

> There was a big gang called the Wilhelmsburger *Türken* Boys. Yeah, that was a group of 200 boys, youth, they met up regularly and fought with some Nazi group or another. They also did lots of criminal stuff, of course. They once attacked the police station here. That was really, really heavy. That was 30 years ago, they disbanded three years later, but it was all – in principle it still has this impact on Wilhelmsburg's reputation. Its reputation was simply completely damaged. It *was* also really bad. As a foreigner in Wilhelmsburg it was totally relaxed, but as a German in Wilhelmsburg it was not so relaxed (Mohammed, interview, 2017).

Though the brief rise of youth gangs in Wilhelmsburg was over three decades ago, and at least 20 years before the state targeted Wilhelmsburg for an image makeover, Yücel argued that it was understandable if people of a certain age still looked at Wilhelmsburg "skeptically."

At the same time, Yücel and Mohammed's descriptions of that period in Wilhelmsburg demonstrate that it was devastating for young, racialized men and for the community more broadly. Since I had not heard about gangs as a current concern in the neighbourhood, I asked Yücel what had changed. He told me, "Yeah, it *had* to change... Because many from [my] generation were deported, and many are dead, deceased, from one thing or another, from drugs and such, and many landed in jail. That was mainly Turkish youth, but also other nationalities. Deported, dead, or in jail. And that was ... that generation didn't survive it" (Yücel, interview, 2017). What he describes strikes me as a human catastrophe that is likely connected to the impression of "zero support" being given the neighbourhood, as Arzu mentioned above. The generation of young people that "didn't survive" would have largely been kids who had come to join their parents who had immigrated in the so-called "guest worker" program, which I will discuss below. They arrived into a hostile country that provided minimal support for "integration." The results were intimately understandable to those who had grown up in Wilhelmsburg themselves. "Not to excuse it," said Mohammed, "but some of them probably felt like I feel [a rejection from German society and exclusion from belonging], maybe worse, and had no prospects." He imagined that they probably thought: "'Fine, if you don't want us, then we'll fight back'" (Mohammed, interview, 2017).

While some Wilhelmsburgers thus argued that there were real conditions that informed the island's image in Hamburg, others argued that its bad reputation has often exceeded the real issues. This is, in particular, due to dominant discourses and representation of "migrants" (actual immigrants and people who are imagined to be immigrants) as problematic, criminal, and dangerous. Bringing us a decade closer to the present, Serdar, who has lived in Wilhelmsburg for 20 years, and is the head of a small local non-profit, remembers that:

> It used to be that this neighbourhood had a very bad image in the '90s. There were also a lot of foreigners here. I mean, historically there were a lot of foreigners here, and in the '90s many, many young Germans left or sent their kids to other schools, to other parts of town. And more migrants came. This majority, or this trend, was portrayed differently by the press and politically, though it wasn't correct. Many always spoke negatively about Wilhelmsburg, though it isn't the real situation (Serdar, interview, 2017).

In Serdar's estimation, media and political representations of Wilhelmsburg incorrectly suggested that migrants were segregating themselves and producing a "bad" neighbourhood, when the reality was that white, middle-class Germans were fleeing the island, as I will discuss in Chapter 5. When I pressed him to say more about what the gap was between the negative image and the real situation, he argued that the media and politicians tended to give the island a hard time, only seeing a small slice of life there and exaggerating dramatically. "For example," Serdar said, "when a crime takes place here in Wilhelmsburg, it is headlines. It is always headlines." Gülhan, a counsellor who was born and raised in Wilhelmsburg, likewise argued that Hamburgers who otherwise had no experience of the island tended to pick up their opinions from the news. These often mobilized the worst of racializing and anti-immigrant representations:

> They see that this neighbourhood has a lot of migrants. And then there is also criminality, and unemployment, and that tends to be pinned on migrants. Migrants suffer because of that, although I don't have the feeling that in other neighbourhoods, I don't know, Dammtor[4] or whatever, certainly also has the same thing, but because fewer migrants live there, it's not shown that way... In the newspaper, when something happens, skin colour is immediately mentioned, and then migrants, and then underlined (Gülhan, interview, 2017).

This kind of representation, which is part and parcel of the criminalization of migrants and racialized others, tended to adhere to the neighbourhood in general, leading to a racialization of space. Racialization of space means that certain spaces come to be associated with particular racialized groups – in this case with a migrantized other – and naturalized as the correct place for those racialized bodies (James, 2012; Razack, 2002; Teelucksingh, 2007). Racialized spaces are examples of what Edward Said (1979, p. 54) called "imaginative geographies:" mental boundaries between "us" and "them," and "our" territory and "their" territory, that originate in the mind but take on material significance. As Said suggests, rather than a simple spatial distinction between natural, pre-existing groups, imaginative geographies are part of the very constitution of an "us" and a "them." The racialization of space and of bodies is thus a dialectical relationship (Razack, 2002) that communicates racial meaning while also producing it.

An analysis of the various labels that have been pinned on Wilhelmsburg over the past decades supports residents' argument that the negative image has a life of its own. "Problem neighbourhood," "Bronx of the North," "social hotspot:" these are all stigmatizing concepts that depend centrally on notions of the "foreignness," danger, and problematic nature of residents. "The Bronx of the North" depends upon the concept of "the ghetto" for its power. As I discussed in Chapter 1, the ghetto has historically signified a part of the city to which racialized people are confined. The concept has travelled from Europe to the United States and back again, and today is taken up as a cautionary racist discourse about the supposed

4 An area of central Hamburg where the main University of Hamburg campus is located.

danger of immigrant self-segregation (Stehle, 2006). Drawing on its tropes of violence, lawlessness and social decay, references to the ghetto communicate fear and desire to control racialized people in and through the neighbourhoods in which they live (Stehle, 2006).

The "problem neighbourhood" (*Problemviertel*) discourse that has been applied to Wilhelmsburg functions in a similar way, as I also noted in Chapter 1. The ghetto and the problem neighbourhood are frequent companions (Tsianos, 2013), mobilized from outside a neighbourhood, and producing a shorthand for people who do not live there. Its use obscures differences within the neighbourhood as well as actual experiences of living there; a problem neighbourhood is simply "wicked, dangerous, and foreign" (Keller, 2015, translation by author). Neighbourhoods acquire the label "problem neighbourhood" based on the presence of poverty, racialization, and low-quality housing (Keller, 2015), or simply on the arrival of immigrants and departure of German families from the neighbourhood (Tsianos, 2013). Both concepts reflect the tendency to blame marginalized people themselves for conditions that are structural, as I will discuss in depth in the following chapters. Wilhelmsburg's bad reputation would thus focus on the threatening history of a Turkish gang, not on the existence of Nazi groups, not on discrimination in work, school and housing, but on "weaknesses" framed as cultural.

This is the case in Wilhelmsburg, where the perceived departure of "German" families (which Serdar flagged, and to which I will return in the following chapters) alongside the impossibility of many families "with migration backgrounds" ever becoming properly German, eventually sparked recent state policies and projects. In Wilhelmsburg it is also evident that the labels attached to the neighbourhood, and the neighbourhood's outwardly bad reputation overall both depends upon and reproduces a sense of it as non-German space. A story that Özden related illuminates how this is the case. When Özden worked at the airport, a colleague of his was invited to visit someone on the island but expressed trepidation about whether as a German he could safely travel there:

> He said: "Can I go to Wilhelmsburg? Can I drive there?" I said: "What do you mean by that?" "Yeah, is it dangerous?" Look: a German didn't want to go to Wilhelmsburg at that time, I'm talking about early 2001, 2002. Not so long ago. He was afraid to go to Wilhelmsburg... So I practically died laughing, then I said: "Wilhelmsburg is still part of Germany, and there are German police in droves. If something happens, you can get help, you don't need to worry" (Özden, interview, 2017).

For Özden, his workmate's fear was both comical and ridiculous. It reflected the probable source of the man's knowledge about the neighbourhood: the local and national media. This anecdote took place in the years directly after the "Bronx of the North" discourse took off in relation to Wilhelmsburg, after several widely reported violent crimes in summer 2000 put the neighbourhood on the national map in a spectacularly negative fashion.[5] The national magazine *Der Spiegel* ran multiple stories about Wilhelmsburg that year for example, in which the island was characterized as a nightmarish, hopeless place full of guest workers and refugees. In one *Spiegel* article, journalist Klaus Brinkbäumer (2000) gave a brief history of "the ghetto," reproducing what I have identified as the dominant narrative about the island's devaluation:

> Wilhelmsburg is the Bronx of the North. The history of the ghetto began in 1962, when the great flood killed 207 people here. At that time a lot of Wilhelmsburgers moved away, and guest workers were put into their dilapidated houses. When new guest workers arrived, the concrete towers were built. Today, 46,000 people live in the backyard of the Hanseatic city; they live between industrial areas, dioxin-contaminated garbage mountains and highways; they live in high-rises without hope, because the atmosphere in Wilhelmsburg breaks everyone down (Brinkbäumer, 2000, translation by author).

This was followed a few months later by an article by journalist Ariane Barth (2000), again in *Der Spiegel*, which called Wilhelmsburg the Balkans of the North, a reference to the presence of refugees from the war in the former Yugoslavia. In the article Barth profiled various Wilhelmsburgers, to some extent capturing its diversity in sections that began with the repeated phrase "Wilhelmsburg is..." Yet the article centred on the threat of a non-German (particularly Turkish) "parallel society" resulting from the perceived characteristics of Wilhelmsburgers with roots in Turkey. At every turn Barth's (2000, translation by author) interviewees claim to have "nothing against Turks," yet bemoan a dwindling "German" population and the isolation of white German children who are the "only ones" in their classrooms. The weekly market is described as an "Oriental bazar," and the practices of Turkish-owned businesses and their customers are portrayed as fundamentally contrary to German standards and normal uses of space. With such characterizations of Wilhelmsburg in wide circulation, it was easy for Hamburgers like Özden's workmate to have never been to Wilhelmsburg and yet already perceive it as distinct, and particularly as non-German and hostile to Germans. Özden's response reflects his own analysis of how Wilhelmsburgers tend to be excluded from Germanness, particularly on the basis of religion.

The perception of Wilhelmsburg as non-German has for decades been specifically wrapped up with the eviction of Muslims from Europeanness that picked up in the 1990s in Germany and consolidated after September 11, 2001 (Ramm, 2010;

5 The crime rate in Wilhelmsburg was in fact lower than in Hamburg in general (Zukunftskonferenz Wilhelmsburg, 2002, p. 2), illustrating that the production of racial fear has a tenuous relationship to facts (cf. Hall et al., 1978).

Razack, 2008). The conflation of the categories of immigrant, Turkish, and Muslim came to dominate German discourse in the early 2000s, and informed perceptions of the neighbourhood (Ramm, 2010; Yildiz, 2009). The tendency to conflate "immigrant" and "Turkish" to begin with reflects the ongoing treatment of people with a background in Turkey as outsiders within, despite their 60-year history in German cities (Çalışkan, 2011; Hinze, 2013). "Turkish" itself often also appears as a unitary category, flattening out significant differences among people with roots in Turkey, and thus produces a homogenous group in contradistinction to the "German" (Çalışkan, 2011). Added to this was Islamophobia, which characterizes "the Muslim" in terms of apparently universal and inherent qualities (Razack, 2008). Though it has never factually been the case that all immigrants and migrantized people in Wilhelmsburg were Muslim,[6] Wilhelmsburg's "bad reputation" often seemed to suggest this, in keeping with the dominant anti-Muslim, racist discourse. The discourse in post-9/11 Europe of the Muslim as terrorist also took on a particular valence in Wilhelmsburg, as Mohammed Atta briefly lived on the island before he flew a plane into the north tower of the World Trade Center (Lauterbach, 2011; McDermott, 2002). I found that his name still came up, particularly in relation to the perceived need for vigilance regarding the threat of (Islamic) radicalization.

In the dominant discourse "the Muslim" (a figure rather than a real person) is imagined as fundamentally outside of European civilization (Ramm, 2010), and as living in neighbourhoods fashioned into archaic and anti-modern societies (El-Tayeb, 2016). This racist and Islamophobic narrative fuels what has been called a "moral panic" (Haritaworn, 2015; Münch, 2009) or an "urban panic" (Tsianos, 2013) in relation to the supposed self-segregation of thus stigmatized communities. Vassilis Tsianos (2013), writing about St. Georg, another racialized and stigmatized neighbourhood in central Hamburg, suggests that this fuels "anti-Muslim urbanism." In anti-Muslim urbanism, discrimination against Muslims and against people perceived to be Muslims, gets justified through references to the unique threat that the Muslim – read as male, straight, and cis-gender – is thought to pose to women and to LGBTQ people in urban space. This is a narrative that has been thoroughly deconstructed by Tsianos (2013) in Hamburg, and by queer of colour and critical race feminist urban scholarship on Northern European cities including Berlin, Amsterdam, Paris, and London (Bacchetta et al., 2015; El-Tayeb, 2012; Haritaworn, 2015; Razack, 2008; Yildiz, 2009).

Anti-Muslim urbanism, the racialization of space, and the racialized stigmatization of Wilhelmsburg through media discourse intertwine with the intentional, structural, and systematic processes of devaluation through planning to create and reproduce material effects on the island. Wilhelmsburg's bad reputation, its status

6 A 2002 report suggests that of 15,700 "foreigners" in Wilhelmsburg, 9,700 were Muslim, but the source for these numbers is not cited (Zukunftskonferenz Wilhelmsburg, 2002, p. 130).

as a racialized "problem," played a major role in its targeting for recent policies and projects. I will introduce these interventions in Chapter 5 and give examples of how the city-state has taken up the problem neighbourhood, ghettoization, and moral panic to justify its plans. In this section, I have introduced the dominant narratives about how Wilhelmsburg came to have a negative image that required state attention. I added to the narratives by highlighting post-flood disinvestment, identifying it as environmental racism and injustice and the deliberate production of a space for waste and work. I then demonstrated how the racialized residents I interviewed saw the neighbourhood's bad reputation as fundamentally based in who lives there and how they are viewed in the prevailing socio-economic order. In the following section I will discuss this order as racial capitalism, before then tracing the racialization of the island further back in time. Residents argued that Wilhelmsburg's bad reputation is critical to its current story, but shifted the state version of it as natural and accidental. I situated residents' analyses within theories about the racialization of Muslims and the ghetto discourse as the definition of non-European, non-German urban space.

In the following section I situate the racialized devaluation of Wilhelmsburg within the city of Hamburg and its historical investments in the production of some people as worth less than others, as a city enriched and shaped by colonialism (Adjei, in Della et al., 2018). The production of differential value is a core process in racial capitalism, which *requires* the racialization and devaluation of some land and labour for the production of value, and produces environmental racism as one of its everyday by-products (Pulido, 2017). The objective of this section is thus an historicization of Hamburg as a city shaped by the logics of racial capitalism, particularly through the dominance of its port, and its influence on the definition of the city. In the final section I will conclude with a brief history of Wilhelmsburg as a space of racialized devaluation, which dates back much farther than the flood narrative would suggest. The history specifically of devalued migrant labour dates back over 150 years to the earliest choices by Hamburg's powerful to treat Wilhelmsburg as a space of waste and work.

Contextualizing Wilhelmsburg within Hamburg, "Germany's number one colonial metropolis"

> "For Hanseatic people this was never the place to live. This is where money was made"
> *(Zukunftskonferenz Wilhelmsburg, 2002, p. 5, translation by author).*

When in 1920 Fritz Schumacher declared the Elbe Islands a space for work rather than for living, 33,000 people were in fact already living in Wilhelmsburg (Behörde für Schule und Berufsbildung, n.d.). The population was almost exactly the same proportion of the city's total as it is today. However, Wilhelmsburg had by that time already become a significant immigrant neighbourhood in Hamburg, comprised largely of workers who were not considered to be citizens of Germany, let alone the proper subjects of urban planning. As the site of industrial and port-related work, performing a crucial function in the life of the city as a whole and in the production of wealth for a certain segment of the population, Wilhelmsburg was devalued in order to produce profit and power within racial capitalism.

In *Black Marxism*, Cedric Robinson (1983) argued that, put simply, capitalism *is* racial capitalism. In order for capital to accumulate, it must move through relations of severe inequality among humans (Melamed, 2015). Capitalism thus requires disposability and unequal differentiation of human value to produce profit. As Melamed argues, "racism enshrines the inequalities that capitalism requires... displacing the uneven life chances that are inescapably part of capitalist social relations onto fictions of differing human capacities, historically race" (Melamed, 2015, p. 77). As Jodi Melamed (2015, p. 82) puts it in her exploration of the concept's use to name and analyze the production of social separateness, "where capital accrual exists, the diminishment of social well-being through partition, dispossession, and appropriation has already happened." The concept of racial capitalism thus advocates attention to the origins of capitalist wealth, and to the role played by colonialism, imperialism, and enslavement in these origins. Racial capitalism demonstrates that we are still living with the legacy of these processes (Melamed, 2015; Pulido, 2017).

In an application of Robinson's concept, geographer Laura Pulido (2017) further argues that racial capitalism produces environmental racism as part of its normal, everyday functioning. The devaluation of racialized bodies that capitalism requires is produced through the appropriation of land and control of access to land, and through racialized systems of labour that sort bodies and produce some that are surplus (Pulido, 2017). Out of this devaluation and extraction of value comes industrial pollution and the restructuring of nature; people who are racialized as non-white disproportionately bear the burden. Pulido asserts that this is not due to regulatory failure, but to devaluation by the state (Pulido, 2017).

I take the presence of environmental racism in Wilhelmsburg as evidence that the processes of racial capitalism are significant to how Wilhelmsburg has been structured and viewed. Wilhelmsburg has been defined as a space of racialized labour since the late 1800s, and thus has a devalued but crucial role in the production of profit, particularly for people to the north of the Elbe river (and beyond). This role in capitalist production was consistent with Hamburg's driving interests as a colonial metropolis. The coloniality of Hamburg has been under-recognized in urban research and history of Hamburg until recently, when Black and People of Colour organizing and scholarship in the city produced some attention to its colonial legacy. Together with the long history of immigration in Wilhelmsburg, to which a number of Wilhelmsburgers encouraged me to attend in my research, this suggests the centrality of racial capitalism to the island's development.

Hamburg: Gateway to the world

Hamburg is a city of great wealth, the historical foundations of which are dramatically illustrated in its city hall. Built in 1895 at the culmination of several decades of port expansion to accommodate overseas shipping, the *Rathaus* building physically and symbolically combines political and economic power that had long been intertwined in practice; it houses the seat of government, chamber of commerce, and stock exchange under one roof (Meyer-Lenz, 2016). Each room in the city hall is decorated more elaborately than the last, displaying the riches and access to materials accumulated by Hamburg's powerful through maritime trade. Multi-coloured marble, wood inlay, stamped and painted wallpaper, and intricately worked metals; the decoration of the city hall makes a statement about the power of Hamburg in general, and especially of its merchants. In the hall's reception rooms, paintings and sculpture depict the city's port history and its treasured trading relationships.

Many of Hamburg's trade relationships were deeply unequal and shaped by the colonial logic of some people being worth more than others (Adjei, in Della et al., 2018). Hamburg has been called Germany's "number one colonial metropolis" (Zimmerer, in Norddeutscher Rundfunk, 2018, translation by author) because its investments in colonial trade and governance are long-standing, not only in relation to German colonialism, but to the colonialism of other European powers. These investments made Hamburg the city it is today in terms of wealth and in terms of its relationship to its "others" (Della, in Della et al., 2018; Seukwa, in Schepers, 2018). Black and People of Colour activists and scholars in Hamburg have been arguing for decades that the city has a colonial legacy, which is present in the space and everyday life of the city, and which it must confront (Della, in Della et al., 2018). This activism has had some significant successes in recent years, leading to the official acknowledgment that there *is* a colonial legacy, and dedication of resources to producing knowledge about colonial Hamburg, particularly in relation to Germany's former colonies (cf. Arbeitskreis Hamburg Postkolonial, n.d.; Behörde für Kultur und Medien, n.d.; Forschungsstelle Hamburgs (post-) koloniales Erbe/ Hamburg und die frühe Globalisierung, n.d.).

The history of Hamburg's port particularly embodies the city's long-standing investments in racial capitalism and colonialism. Though Hamburg's colonial legacy is rooted in various institutions, the port is central to the city's identity and to its urban planning priorities, and is considered the economic foundation of Hamburg (von Beust, 2005). Indeed, historians of Hamburg argue that for centuries urban development and port development were fundamentally the same thing; the city-state always tried to anticipate and respond to the needs of the port (Meyer-Lenz, 2016). In the face of the interests of merchants, ship owners, local exchanges, insurance companies, and other key port-driven industries, other interests tended to be sidelined (Meyer-Lenz, 2016). As former mayor Ole von Beust (2005, p. 13, translation by author) put it in a speech about new development strategies in Wilhelmsburg: "To date, the philosophy of urban development has been that the port area is a kind of shrine, the 'eighth district.' Hamburg is divided into seven districts, and the vast port area is the eighth district because port and economy have always been priorities. After all, Hamburg lives very well with and from its harbour."

In an illuminating variation on the idea of the port as "sacred," I several times heard the port referred to by Wilhelmsburg activists as the city's "golden calf." From this I gathered that the rule of development in Hamburg was that the port is always the top priority, and, at minimum, no development can disrupt or be perceived to undermine the port. In an interview, a local planner described the Hamburg Port Authority to me as the most powerful actor in Hamburg, "like a state within the state" (interview, June 6, 2018). The photo below (Figure 12) was taken at dusk in 2011 on my daily bike ride between St. Pauli and Wilhelmsburg, on which I passed through part of the port lands and through the Old Elbe Tunnel to the other side of the river. The photo shows "The Golden Calf" by Elisabeth Richnow, which was installed from 2008-2013 on a pillar rented from the Hamburg Port Authority, at a spot where residential Wilhelmsburg ends and the port lands begin. As Richnow describes it: "the golden calf is a quote from Christian mythology, Moses' brother and his followers pray to it. For me it stands for the dance around something false, the worship of wealth and power...Will all these planning projects, that so much money is being pumped into, make people live happily ever after in the city?" (Richnow, quoted in Maeck, 2009, translation by author).

At a moment when new planning policies and projects were being implemented in Wilhelmsburg, as I will describe in Chapter 5, Richnow uses the Biblical symbolism of the worship of a false idol to enquire about what that development will do,

Figure 12: "The Golden Calf" by Elisabeth Richnow illuminated at dusk, Wilhelmsburg

(photo by author, 2011)

and whose interests it will serve. She strategically locates this question at a juncture between the port and other land uses. I was struck by the installation when I saw it, but only returned to the photo when someone used the term "golden calf" years later and my memory was jogged. What I notice when I look at the photo now is that the golden calf is perfectly framed by port infrastructure: roads and bridges on which containers are carried in and out of the port lands; the stop for the ferry that takes workers to and from work; and in the distance, the cranes that load and unload container ships.

The effects of the "sacredness" of the port have historically been differential. One illustrative example is in the construction of the *Speicherstadt,* the world's largest warehouse district, which began in 1883 and was ultimately completed in 1920. It was constructed within a customs-free zone (the *Freihafen*) that Hamburg had demanded as a condition of joining the German nation-state (Arbeitskreis Hamburg Postkolonial, 2018; HafenCity Hamburg GmbH, n.d.-b). To enable the construction, the city of Hamburg evicted 20,000 people and demolished their housing, based on the aspirations of capital and government; in an "exercise in national imagination" they were hoping for a dramatic expansion of overseas trade that would require more warehousing capacity (Mancheno, in Hengari et al., 2018). Thus part of the city core was emptied of people and rebuilt "not for people but for goods" (Mancheno, in Hengari et al., 2018). Its symbolism remains today, as the *Speicherstadt* has become a shrine of a different kind; obsolete since the containerization of the port, the district is now a UNESCO World Heritage Site and one of the city's top tourist attractions (Freie und Hansestadt Hamburg, n.d.-b).

The aspiration that fueled the construction of the *Speicherstadt* was informed by Hamburg's colonial profits up to that time and by the hope that they would imminently expand and be secured through German colonialism. Hamburg merchants, ship owners, and industrialists had been extracting profit from colonized lands for centuries, through the shipping, trade, and processing of products such as rubber, palm oil, cocoa, and enslaved people (Zimmerer, quoted in Norddeutscher Rundfunk, 2018). Though on paper they were barred for many years from trading at all with the colonies of other European powers, Hamburg companies found many ways to work around this (Todzi, 2018). They sent ships and representatives to register in nearby Altona, which at the time was a Danish city and is now part of Hamburg. In this way they used the port as a stopover for slave ships between West Africa and the Caribbean and profited directly from the abduction and enslavement of people. At the end of the 18th century, Hamburg was also the European centre for sugar cane processing and cotton finishing, having taken on a different – yet central – role in trade they were barred from doing with Hamburg ships (Todzi, 2018). From the mid-19th century, some Hamburg merchants operated plantations in Africa themselves, using slave labour to produce raw materials for trade (Norddeutscher Rundfunk, 2018).

Prominent Hamburg merchants and ship-owners were thus a major force in the push for establishment of German colonies in Africa after unification in 1871 (Norddeutscher Rundfunk, 2018; Todzi, 2017). In 1883 the Hamburg Chamber of Commerce sent a memorandum to the government of the German Reich requesting that it establish "protectorates" in West Africa based on the motto "the flag follows the trade" (Todzi, 2017). The recommendation was that the state should protect German trade interests through the creation of chartered companies and by force when needed. That push was eventually successful; after the "Berlin Conference" in 1884 Germany "acquired protected areas" in Africa, Asia, and the South Pacific (for a map, see Todzi, 2017).

This period of Germany's colonial career was brief compared with that of other European powers, and is said to have ultimately cost the German state financially (Della, in Schepers, 2018). Hamburg, however, turned significant profits, not only from the export of raw materials, but from the provision of ships and supplies that the German military required to get to Africa and enforce German rule (Schepers, 2018). This enforcement was devastating from the outset; in 1884 the Duala in Cameroon refused to recognize German rule and warships were sent to "convince" them (Todzi, 2017). In 1904-1905, the Herero and the Nama in what was called German Southwest Africa (now Namibia) fought back against dispossession and theft by German colonizers. What followed was genocide, in which German troops systematically killed 100,000 people over four years by forcing them into the desert, poisoning their water sources, confining them to camps, and shooting them on sight. Eighty per cent of the Herero population and 70% of Nama were killed (I. Transnational Congress on the Ovaherero and Nama Genocide, 2016; Norddeutscher Rundfunk, 2018).

As a result of the activism of descendants of the victims of the genocide, and of Black people and People of Colour in Hamburg, the city has been forced to face its role as a transport centre for the tools of genocide. Crucial supplies in the systematic effort to destroy Herero and Nama people, including troops, weapons, horses, and poison to contaminate water sources, were all shipped from Hamburg to the colonial authority in German Southwest Africa (Hengari, in Hengari et al., 2018). Some Hamburg companies, such as the Woermann shipping company, share direct responsibility for the genocide, as without their services it could not have happened. Many other companies and individuals profited less directly but nonetheless significantly, such as suppliers and outfitters of the ships themselves (Zimmerer, in Schepers, 2018). In 2018, Carsten Brosda, the Hamburg Senator for Culture and Media, apologized to attendees of the Second Ovaherero and Nama Conference in Hamburg for the city's role in the genocide (Brosda, 2018). The apology was received by at least some descendants as a genuine and respectful step in the right direction (Kamaţuka, in Della et al., 2018; Hengari, in Hengari et al., 2018).

Yet the material and discursive vestiges of colonialism are still very much present in Hamburg. They are visible in the cityscape, in monuments and buildings that honour colonial criminals and that idealize colonial relations. These are not exclusively *old* traces, as in the case of the likeness of Lothar von Trotha that was built into and still stands on the front of the army barracks in Hamburg-Jenfeld.[7] Descendants of genocide victims have specific suggestions about how to address these (cf. OvaHerero, Mbanderu and Nama Genocides Institute, 2018). There are also multiple recent examples of the attempted restoration of colonial monuments that have been questioned and rethought as a result of the activism of Hamburg's Black and African community. These include so-called Tanzania Park, in which reliefs depicting African soldiers and bearers in "German East Africa" were restored and remounted with public funds but without appropriate, critical contextualization (*Afrika-Hamburg.de*, n.d.; Arbeitskreis Hamburg Postkolonial, n.d.; Schepers, 2018). Herero scholar and activist Ngondi Kamaţuka argues that these are evidence of the city's willingness to cause pain to descendants of the victims of colonial crimes, and that it does not yet hold itself fully accountable

7 Von Trotha was the commander who gave the order to "destroy" the Herero and Nama, and who famously wrote that he would crush their resistance "with rivers of blood and rivers of money" (cf. Schepers, 2018).

for its role in those crimes (Kamaṱuka, in Della et al., 2018; The Association of the Ovaherero Genocide in the USA, 2017).

At the same time the city produces *new* monuments to colonialism that reflect a similar willingness on the part of the city. HafenCity is a recent redevelopment of former port lands into an entirely new neighbourhood mainly for the rich (see HafenCity Hamburg GmbH, n.d.-a). There are new buildings, squares, and bridges that are named after plantations, colonial products, and European "explorers" and colonizers (Mancheno, 2016). The naming is meant, not unlike the decoration of the *Rathaus*, to evoke the historical and present day "overseas" connections of Hamburg, and to strike a worldly tone for Hamburg's urban development. A monument to coffee in a public square, for example, evokes images of HafenCity dwellers as civilized and cosmopolitan through the consumption of colonial products (Arbeitskreis Hamburg Postkolonial, 2018). But the naming choices depend upon what has been called Hamburg's "colonial amnesia:" inattention to how those worldly references are signs and symbols of death, displacement, and dispossession (Kellermann, 2019, drawing on Mancheno; Mancheno, 2016). Colonialism does not explicitly feature at all in HafenCity's account of the "historical ground" on which it stands (HafenCity Hamburg GmbH, n.d.-b), and the result is what local scholar Tania Mancheno calls an "echo that is based on colonial power, and that does not remotely consider the racism that was justified by its adoption" (quoted in Kellermann, 2019). This "echo" is possible because the structures of colonialism live on in Hamburg through racial capitalism, and through how some people are respected and valued more than others along racial lines (Adjei, in Della et al., 2018; Choukri, Gläser & Prinzleve, in Forschungsstelle Hamburgs (post-) koloniales Erbe, 2018). Such echoes contribute to the reproduction of the city as belonging to some and not to others, in keeping with the coloniality of the European city in general, which is built from colonial extraction and exploitation and which persists in the colonial control of people from the global South (N. Ha, 2014a). As Mbakumua Hengari (in Hengari et al., 2018) puts it, colonialism is like a mega-project: the infrastructure built for it persists well beyond the nominal "end" of the project, and the benefits and the costs continue to accrue. The organizing of Black and People of Colour Hamburgers against the city's colonial amnesia chips away at this infrastructure and seeks to redistribute the benefits and the costs (cf. *Afrika-Hamburg.de*, n.d.; *Colonial Amnesia*, n.d.).

The focus of current critical scholarship and activism is mainly on the traces of Hamburg's overseas colonial ties, as a result of the city's specific maritime history and the urgency of challenging the exceptionalism that had long situated Germany's colonialism as less important than that of other European powers. Yet Germany also has a history of colonial expansion in Europe, and I will demonstrate in the following section that the racialization of people from what is now Poland, and themes of anti-Slavic racism, have played important roles in Wilhelmsburg's history. German national formation in the 19th century, and invasion of neighbouring countries in the 20th century, were pursued in often overtly colonial terms that framed people and spaces as under-developed and thus in need of German settlement and governance (Kopp, 2010). In 2021, Hamburg held an online public workshop to discuss the "contextualization" for example of the massive, prominent statue of Otto von Bismarck that overlooks Landungsbrücken (Behörde für Kultur und Medien, 2021). Concerns were raised by the public about continuing to dramatically valorize the originator of the Berlin Conference at which European powers carved up the map of Africa (Todzi, 2020; Iken, 2020). Bismarck also pursued a campaign of what was called "inner colonization" of Prussia, moving "ethnic Germans" into Polish-speaking areas to secure German control in racial terms. This expansion would later be revisited under National Socialism as a reclamation of "German soil" (*Kulturboden*), imagined as decaying under "Slavic" control, and as a move to "defend" Germany from the uncivilized and threatening East (Kopp, 2010).

At the core of this colonial history is again the structure in which some people are considered to be worth less than others, which is used to justify colonization, exploitation, and enslavement, often in the name of "civilization." Black and POC scholars and activists in Hamburg have centred this logic, in order to analyze how the vestiges of colonialism continue to play out in the city, not only for the descendants of those formerly colonized by Germany, but for a range of people who are differentially racialized according to interconnected logics.

Racialized inequality in the city is one striking vestige. Currently, Hamburg has the highest per capita income in Germany and the most millionaires per capita (Handelskammer Hamburg, n.d.). In 2017, 867 Hamburgers had an income of a million euros or more (Statistisches Amt für Hamburg und Schleswig-Holstein, 2017). In 2018 the city saw record sums change hands for residential real estate, with houses going for as much as €7.5 million in the wealthiest neighbourhoods (Preißler, 2019). Yet it is also a city of racialized economic inequality; there is a major income gap between the richest and the poorest, which is expressed in dramatic concentrations of wealth and poverty across the city. Hamburg frequently ranks highest in Germany in levels of economic segregation, and low-income households are generally being pushed out of the inner city (Güntner, 2013). The average income of the city's poorest neighbourhood is less than 11% of the richest, and the poorest neighbourhoods tend also to be those with the highest percentage of racialized residents (see Statistisches Amt für Hamburg und Schleswig-Holstein, 2017, p. 5, together with 2018, p. 12). This situation has to be read across statistical reports, however, as the identification of racialization and racism as factors shaping Hamburg neighbourhoods is very rare in the local context except among Black and People of Colour scholars. Instead the focus tends mainly to be income and class stratification and polarization, with occasional acknowledgment that the presence of poverty and immigrants and/or "people with migration backgrounds" overlaps.

However, as Sherene Razack argues, "when affluent areas of the city are all white and poorer areas are mostly of colour, we experience the spatiality of the racial order in which we live," (Razack, 2002, p. 6). In the following section I continue to explore the racial order as it manifests in Wilhelmsburg in particular, specifically in the island's long history of immigration. A century before the 1962 flood Wilhelmsburg was already being produced as a racialized space through the deliberate location of devalued people and devalued city functions on the island. The island has a deep history not only of immigration in general, but of labour migration in which workers were recruited specifically because they could be treated differentially: paid less and provided poorer working conditions than the state was willing to allow for its citizens. What I have demonstrated in this section, through a short history of Hamburg's colonial entanglements, is that this treatment of racialized others is consistent with Hamburg's history, rooted in the city's investment in colonial production of profit and wealth, which has depended upon the racialization, devaluation and indeed murder and enslavement of people constructed as "others." To the extent that Hamburg is a port city and a city with a colonial legacy, the wealth that shapes and influences it is based on assumptions about inequality in the production of value. In Wilhelmsburg this has been reflected in long-standing environmental racism, as well as the recruitment and differential treatment of various immigrant groups.

Wilhelmsburg's long history of immigration

Wilhelmsburg's immigrant population grew alongside industrialization and expansion of Hamburg's port. The island itself was born in something resembling today's form when it was diked out of smaller islands in the 17th century, and named after the then Duke of Lüneburg-Celle, Georg-Wilhelm (Geschichtswerkstatt Wilhelmsburg und Hafen, n.d.-a). Until the mid-19th century, the island was exclusively farmland, and it had a very small population that produced fruit, vegetables and dairy for the local market (Geschichtswerkstatt Wilhelmsburg und Hafen, 2018). The local Sinti community dates its presence on the island to around this time (Journalisten der Henri-Nannen-Schule, 2015). In the late 19th century, the island drew the attention of land speculators in Hamburg as the city anticipated desire for industrial development and found in Wilhelmsburg the space that it would demand (Geschichtswerkstatt Wilhelmsburg und Hafen, n.d.-b). At the same time, Hamburg was engaged in its aspirational expansion of overseas trade, including a customs-free area (*Freihafen*) that the city-state demanded as a condition of joining the German customs union (Meyer-Lenz, 2016; Spiegel Online, 2012). The *Spei-*

cherstadt, whose development I described above, was built within this zone. As the port expanded, so did shipbuilding and oil industries in Wilhelmsburg, and by 1891 prominent Prussian officials were calling Wilhelmsburg the "most ideal industrial area in the German Empire!" (Geschichtswerkstatt Wilhelmsburg und Hafen, 2008, p. 2, translation by author).

The devaluation of racialized labour was a defining feature of industry and thus of life in Wilhelmsburg from those earliest years. Hamburg businessmen actively recruited Polish-speaking peasants in Prussian-occupied Posen, and the people who came were driven by deep poverty, marginalization, and unemployment in what is now Poland (Geschichtswerkstatt Wilhelmsburg und Hafen, 2008, 2018). Kien Nghi Ha (2007a, p. 66, translation by author) has called this a kind of "inverted colonial expansion" or "internal colonization," as people were brought to work in western Germany while German subjects were encouraged to settle in the Polish-speaking east. Once in Wilhelmsburg Polish-speakers faced dire conditions, as they were recruited to do the most difficult and dangerous work for starvation wages. This was often skilled work that was devalued as menial or as "women's work" (Geschichtswerkstatt Wilhelmsburg und Hafen, 2018).

The racialization of Polish-speaking workers at the time is evident both in the local historical record and in the labour migration policy of the German Reich. "Most Wilhelmsburgers [looked] down on them" (Geschichtswerkstatt Wilhelmsburg und Hafen, 2018, p. 27, translation by author), and complained in the local newspapers about Polish-speakers being thieves and alcoholics. They were considered fundamentally different because of their language, religion (Catholicism), and lack of education (Geschichtswerkstatt Wilhelmsburg und Hafen, 2008). This was the case beyond Wilhelmsburg as well, as at the time Polish workers were generally racialized as inherently less intelligent than Germans and as natural manual labourers (K. N. Ha, 2007b). This racialization was reflected in German labour policy, which claimed that the country recruited foreign workers to do what Germans were "too culturally advanced" to be interested in, and that this was a healthy and "hygienic" means of in fact developing inherently different peoples (Geschichtswerkstatt Wilhelmsburg und Hafen, 2008, p. 4). Yet labour migration policy was also explicitly planned to maximize German profits by developing a workforce so precarious that it would be flexible to the whims of German industry. In contrast to a workforce of citizens, who might make demands if they were unemployed and hungry, foreign workers "could simply be discarded" in the case of any industrial decline (Prussian ministry of trade, quoted in Geschichtswerkstatt Wilhelmsburg und Hafen, 2008, p. 5, translation by author). Around the turn of the century 30% of workers in Wilhelmsburg were considered disposable in this way, because they were not German citizens, because they were racialized, or both.[8]

The wool-processing factory, meticulously documented by local historians, provides a stark example of how this functioned in Wilhelmsburg. Founded in 1889, the factory employed 2,000 people, most of them women, and almost exclusively Polish immigrants. The work of the factory was to process raw wool that was brought off ships in huge rolls, probably from Australia and New Zealand (M. Markert, personal communication, December 19, 2018). Twelve-hour shifts were worked in deafeningly loud conditions, with a 15-minute break, and wages were the worst in all of Wilhelmsburg, despite the risk of injury. Workers struck for better wages in 1906, but were successful mainly in building solidarity amongst workers in Wilhelmsburg. When workers went home at the end of the day it was to extremely small, dark, damp, and cold apartments that were hastily constructed by their employer and tied to their work contracts: If you left or lost one, you also lost the other (Geschichtswerkstatt Wilhelmsburg und Hafen, 2008, 2018).

The history of Polish-speaking people in Wilhelmsburg remains part of the social and spatial fabric of the island today, which I believe is a reason why I encountered more than one Wilhelmsburger who urged me to attend to the *long history* of Wilhelmsburg as an immigrant neighbourhood. The early industries are reflected in street names such as Bei der Wollkämmerei, and at the base of Veringstrasse you can still attend Bonifatiuskirche, a Catholic church that was built by the Polish community in 1895. Racialization shifts and changes over time (cf. Allen, 2012), and Germany's relationship to what is now Poland also shifted and changed, while the colonial discourse of "under-development" and the German as a "civilizing" force continued to appear (Kopp, 2010). In the 20th century, Germany invaded and occupied Poland twice, the second time aiming to eradicate Slavic people and cultures and create "*Lebensraum*" for Germans, turning Poland into a "slave colony," and ultimately killing 20% of its population (Kauffman, 2015; Niewyk & Nicosia, 2012). Yet the Wilhelmsburg History Workshop suggests that over time Polish-speakers were no longer judged to be outsiders on the island, and today their descendants consider themselves to be Wilhelmsburg's "natives" (Geschichtswerkstatt Wilhelmsburg und Hafen, 2018).

From "foreign workers" to "guest workers"

The long history of immigration to Wilhelmsburg, and the specific devaluation and exploitation of racialized workers who came to identify *as Wilhelmsburgers* is important to understanding the island and its stigmatization because it is part of a

8 Some Polish speakers were officially citizens of the German Reich though nonetheless racialized, while others were not.

pattern. The descriptions of working and living conditions for immigrants and the logic of German labour-migration policy could easily have been written 70 years later, about the migration of the so-called "guest workers." The "guest worker program" was a German immigration and labour policy from 1955 to 1973. Through agreements with countries in Southern Europe, West Asia, and North Africa, migrant workers were recruited, particularly from impoverished areas, to do dangerous and dirty work in German industries (Chin, 2007, 2009). There was, again, a lot of this work in Wilhelmsburg's factories and in the neighbouring portlands (Dietz, 2008; Geschichtswerkstatt Wilhelmsburg und Hafen, 2008). Half of the Wilhelmsburg residents I interviewed were personally connected to this migration, as the children, grandchildren, or in-laws of former "guest workers." With the exception of Paula, whose parents are from Spain and Portugal, most of these residents are Turkish speakers, which reflects the fact that the largest group of workers came to Germany from Turkey between 1961 and 1973. Wilhelmsburg's demographics also reflect this today; according to a recent district report, Turkey is at the top of the list of "countries of origin" for Wilhelmsburgers (22.4%), after Germany (43%), and followed distantly by Poland (4.75%) (Bezirksamt Hamburg-Mitte, 2015).

The devaluation of racialized workers was again evident when it came to migrant workers from Southern Europe, North Africa, and West Asia, as racialization and objectification continued to be built right into German immigration and labour policies (K. N. Ha, 2003, 2007b). Much of what was considered unacceptable for the average German in terms of work, pay, safety, and housing, was treated as sufficient for so-called guest workers. In Wilhelmsburg, workers were often housed in low-quality, company-built housing, including in barracks (Dietz, 2008). Though the "foreign" populations changed, the fundamental logic of the regime remained consistent: increase profits for German companies by decreasing costs, by bringing in workers who are considered "cheap and willing" (Chin, 2009) and easily disposed of (Chin, 2007, 2009; K. N. Ha, 2003, 2007b). The status of "guest" was tied to real practices of legal and political separation from the German population, subjection to separate laws and employment standards, and lower wage entitlements (Chin, 2009).

This differential status is emblematic of how racial capitalism functions. In order to produce *profit*, here generally for the owners of German companies, certain bodies are racialized and positioned as not as valuable as Germans and thus not entitled to the pay, protections, living conditions and so on, that equality demands, but that would constrain profits. The devaluation of racialized workers produces *power* for the state, and again for German companies at the same time, which are afforded increased control over all workers through the pressure that racialized exploitation puts on wages, working conditions, housing and so on. "Guest worker" policy also illuminates, however, how racism can exceed the desires of capital within racial capitalism. National policy was to avoid recruitment treaties with non-European countries (Turkey was considered a part of Europe at the time), and to actively prevent employers from recruiting workers from Africa and Asia (Schönwälder, 2004). Despite pressure and protest from employers, embassies, and the foreign office, federal ministries aimed to especially stop African and Asian men from settling in the country on the basis of their supposed cultural unsuitability (Schönwälder, 2004). A differential racialization is thus noticeable in the policy, wherein the labour of Southern Europeans, Africans, and Asians was all devalued, but through a calculus of the likelihood of workers returning whence they came, and of culturalized and gendered "risk" to Germany.

In between the two periods I detail here, Germany attempted to "eradicate its minorities" (Chin, 2007, p. 10) including precisely through work. The labour of prisoners of war, prisoners from concentration camps, people deported from Eastern Europe, and people who were imprisoned locally, was used by many companies in Hamburg during World War Two, including in Wilhelmsburg (Landeszentrale für Politische Bildung Hamburg, 2007). Emil Weiss, head of the Wilhelmsburg Sinti community until he passed away in 2018, was as a 13 year-old forced to work in a factory in nearby Harburg (Journalisten der Henri-Nannen-Schule, 2015). Detailed interactive maps produced by researchers at the former main concentration camp in Hamburg, KZ-Neuengamme, illustrate where forced labour took place in particular in war production and how many people were enslaved where. The map shows that among perhaps one hundred sites of forced labour in Wilhelmsburg, 173 people were enslaved at the wool processing factory in Wilhelmsburg, for example, and 102 at the tin factory; closest to where I lived in Old Wilhelmsburg was a camp with 118 "*Ostarbeiter*" ("Eastern workers") at the railway, the Deutsche Reichsbahn (Landeszentrale für Politische Bildung Hamburg, 2007). The Langer Morgen "Work Education Camp" was also located on the outskirts of Wilhelmsburg from 1943-1945. The purpose of the camp was to punish and "retrain" defiant or uncooperative enslaved people, as well as people who were considered anti-social, "work-shy," criminals, quitters, or sex workers. Eighty per cent of the people imprisoned there were non-Germans, and most were from the east (Frank, 1997).

The devaluation of racialized labour continues today. In fact, one of the recent "changes" that I often heard about in Wilhelmsburg was the recent arrival of "the Bulgarians" to the island. "The Bulgarians" refers to actual Bulgarians, as well as Romanians, who began to migrate to Germany after the countries joined the European Union in 2007. Many but not all of "the Bulgarians" are Roma, and many are also part of a Turkish-speaking minority. The latter makes Wilhelmsburg a particularly attractive place to land, as in Wilhelmsburg one can get most basic needs met in Turkish. This can be a major boon for newcomers, as Ayşe, a 20-year resident of Wilhelmsburg put it to me: "A Turkish shop, Turkish doctor, and Turkish – for example at [the supermarket or drugstore] there is Turkish staff. Turkish baker. It is a disadvantage [if you are trying to learn German], but if I don't speak German: advantage" (Ayşe, interview, 2017).

Eastern Europeans are encountering conditions in Wilhelmsburg that again sound like historical descriptions of what Polish-speakers faced in the late 1800s. Low wages, unhealthy and dangerous work, and terrible living conditions are the typical state of affairs (Lintschnig & Schultz, 2018). What has changed from a century ago is the precise structure of exploitation, which now assumes a neoliberal form, in which workers rarely have employment contracts and are instead nominally "self-employed." According to local laws and practices, without an official work contract one cannot get a legal apartment; without an apartment one cannot register with the authorities; and without registration, your kids cannot go to school. The effects of this bind are often blamed upon "the Bulgarians" themselves, who are derided as supposed alcoholics, fighters, thieves, and beggars: precisely the same terms that were used for Polish-speakers a century ago (cf. Geschichtswerkstatt Wilhelmsburg und Hafen, 2008). In particular, this group of labour migrants is criticized for their use of public space, which clashes with bourgeois values regarding how and when one should be in a park, for example. Differences are at times framed by other Wilhelmsburgers as "cultural" qualities belonging to "the Bulgarians," a move which fixes difference (Hall, 1997), and which several of the residents I spoke to flagged as an example of racism among racialized people.[9] I will discuss this in more detail in following chapters, in view of the particular social, economic, and political conditions that inform differential racialization (Pulido, 2006) here.

The system through which labour migrants in Wilhelmsburg are exploited is one from which German companies and consumers profit massively (Lintschnig & Schultz, 2018). This profit and power do not, however, generally accrue to Wilhelmsburgers themselves. I heard some indications in the neighbourhood that there are a few local landlords who are taking advantage of the vulnerability and precarity of newcomers to the neighbourhood by renting spaces not designated for housing, such as cellars, attics, and warehouses, or by renting single rooms to many people for well above market rent. There are also a few local employers who are similarly extracting profit from the extreme devaluation of Bulgarian labour (Lintschnig & Schultz, 2018).

In general, however, Wilhelmsburg is predominately a working-class and low-income neighbourhood and has historically been a bastion of leftist, communist,

---

9 I am grateful to the working group *Sasedi* Wilhelmsburg, who were conducting their own research while I was there into the issues Bulgarians face in Wilhelmsburg. My host and colleague Michael Rothschuh invited me to sit in on group meetings, which I was able to do for several months; as a result I was exposed to many first-hand accounts of the issues from local service providers and from interview transcripts, which enriched and clarified what I heard in my own interviews.

and social democratic politics. This is a way in which its heterogeneity has long had a distinct character within Hamburg. The island was hit hard by the containerization of the port and by deindustrialization and thousands of jobs disappeared (Zukunftskonferenz Wilhelmsburg, 2002). According to Wilhelmsburg's representative in German Parliament, Metin Hakverdi (SPD) (interview, 2018), the island's residents have never fully recovered from this "restructuring;" the kinds of jobs for which people were trained, recruited, and indeed for which many immigrated, simply disappeared and never returned. During this study, Wilhelmsburg had rates of unemployment that were twice as high as the city-wide average, particularly among residents 55-65 years of age and 22.5% of Wilhelmsburgers relied on social assistance to make ends meet, compared with 10.3% city-wide (Statistisches Amt für Hamburg und Schleswig-Holstein, 2018, p. 48). The average income was among the lowest in the city (Statistisches Amt für Hamburg und Schleswig-Holstein, 2017, 2018).

This is not simply an accident of planning to be corrected by "mix." Rather, Wilhelmsburg has been produced deliberately as a space of devalued work for over a century, and the dominant narratives tend not to capture that. If, however, one approaches the devaluation of Wilhelmsburg as being consistent with the economic and social order, which the city-state has no intention of deeply challenging, then a question arises about what is to be achieved with "mixing" for people who are devalued within that order. What are the recent planning interventions *imagined and intended* to achieve and what might they actually do for Wilhelmsburgers? It is to these and related questions that I turn to in Chapters 5 & 6.

Conclusion

In this chapter I have traced how Wilhelmsburg's "bad reputation" hinges fundamentally on the kind of people who are imagined to live there – immigrants, low-income people, the unemployed, people on social assistance – in short, people who are stigmatized and devalued in Hamburg and in German society. The residents I interviewed argued that the reputation is important to understanding current events, and attributed it to a combination of factors that included actual conditions in which people of low socio-economic status were deliberately concentrated on the island, and racist discourse about the ("dangerous," "criminal") presence of migrants. This perspective does not contradict the dominant narrative, but rather connects the dots within it, drawing attention to the many steps in the process of devaluation of Wilhelmsburg. Situating this devaluation within the city's investments in colonialism and racial capitalism illuminates how particular assumptions about the value of who lives there have been integral to the island's development and reputation since long before the flood that is typically cited as a problematic turning point.

The racialized residents I interviewed see Wilhelmsburg not as a problem neighbourhood, however, but largely as a warm, welcoming, and beautiful place, and I turn to discussion of these perspectives in the following chapter. The past and present of racial capitalism has also produced conviviality in the neighbourhood, and Wilhelmsburg is *Heimat* for many of its long-time residents, a place of belonging and comfort where racialized people and migrants feel at home, in contrast to the exclusions that they experience as racialized people in Germany.

4. Heimat Wilhelmsburg
Belonging and Familiarity in the Context of Devaluation

> I want to say 'yes, I am a Hamburger, though with a migrant background' – but Wilhelmsburg's identity is open: I can say 'I am a Wilhelmsburger', and I'm happy to say it.
> *(Serdar, interview, 2017).*

Introduction

Wilhelmsburgers love their island. While I met the occasional resident who said they would consider moving elsewhere if it were not for a job or children or other practical concerns tying them to the neighbourhood, most of the Wilhelmsburgers I interviewed and those I met during everyday life were strikingly fond of where they live. At first, I wondered whether this was a manifestation of selection bias, wherein the people who were interested in talking to me about Wilhelmsburg were those who were passionate about it. When I shared this observation, however, and asked other residents whether they noticed this too, I was often met with smiles and laughter. *Yes*, long-time residents have a special bond to this place.

Through the perspectives of racialized residents, Wilhelmsburg appears to be quite a different place than the supposed "problem neighbourhood" that I explored in the previous chapter. It emerges as a warm, welcoming, diverse, and beautiful place, full of strengths not in spite of, but because of, its long-time devaluation and production as a racialized and migrant space. Conviviality – processes of cohabitation and interaction that have made multiculture an ordinary part of life (Gilroy, 2004) – emerged from the island's history over time. This Wilhelmsburg is *Heimat* for many of its long-time residents, a place of belonging and comfort where racialized people and migrants feel at home, in contrast to the everyday and structural exclusions that they experience as racialized people in Germany.

By exploring these themes and focusing on what racialized residents wanted people to know about their home, this chapter thus fills a gap in knowledge about Wilhelmsburg and contextualizes the "social mixing" planning interventions that

I introduce in the following chapter. Based on my in-depth, semi-structured interviews with long-time residents, I sketch out what Wilhelmsburgers love about the island, from its green and beautiful landscape, to its village-like quality and the warmth and friendliness that they experience with neighbours. Wilhelmsburg, residents argue, is not an urban neighbourhood in which people are anonymous, it is a place where you know each other and are recognized. It is also a place that is diverse, where people with various backgrounds and socio-economic statuses live together. Wilhelmsburgers associate this with the neighbourhood's long history as an immigrant neighbourhood, and with its bad reputation. Over time it has become *Heimat*, the place where they feel comfortable, where racialized people and migrants are accepted and reflected in everyday life, and where important symbols of spatial belonging have been achieved in the face of racism and discrimination.[1]

To understand these perspectives, I open with a genealogy of the concept of *Heimat* as a spatial concept of identity and belonging (Blickle, 2002; Boa & Palfreyman, 2000; Eigler & Kugele, 2012) that is commonplace in the German language but untranslatable to English. In the midst of what has been called its "renaissance" (Costadura et al., 2019), *Heimat* is currently hotly debated and contested in Germany as an exclusive concept with a racist history that has recently been evoked on the national stage (Ataman, 2018a, 2018b; Aydemir & Yaghoobifarah, 2019). Wilhelmsburgers take up *Heimat* in its broadest and most open sense, in keeping with their appreciation of conviviality on the island, to indicate it as a place where they belong and feel safe and secure both as individuals and as racialized people. The genealogy of *Heimat* underscores the significance of long-time residents' attachments and affection for Wilhelmsburg in a context where migrant and racialized belonging in general, and the belonging of racialized Wilhelmsburgers in particular, is continuously challenged and questioned.

Locating Heimat

Heimat does not translate neatly into English, though it is a very common word in German.[2] Depending on the context, it might mean home, homeland, homestead, birthplace, hometown or many other similarly related but distinct concepts in English. Generally *Heimat* refers to a place where one feels a sense of security

1 Some elements of this chapter will appear in a forthcoming issue of the *Journal of Race, Ethnicity & the City*, under the title "Heimat Wilhelmsburg: Belonging and resistance in a racialized neighborhood in Germany."

2 Because of its untranslatability into English, I have elected not to translate *Heimat* and to italicize it throughout, as I do other German words.

and belonging that is connected to the human need for safety and for connection to land and place (Boa & Palfreyman, 2000; Costadura et al., 2019; Räthzel, 1994).

According to scholars of the concept and discourse of *Heimat* and its uses in literature, film, and philosophy, *Heimat* dates back at least to the early 1800s when it described a single place to which one had a legal relationship as a resident (Römhild, 2018). The concept's development into one that is emotional and subjective (Blickle, 2002), and that references a sensory and feeling relationship to space (Römhild, 2018), took place in the context of industrialization and urbanization, when it began to be used as a reference point for identification in the midst of societal changes (Boa & Palfreyman, 2000; Eigler, 2014; Eigler & Kugele, 2012). *Heimat* became a nostalgic concept, temporal as well as spatial, as it expressed a longing for a time and place outside of modernity (Eigler, 2014; Eigler & Kugele, 2012), and for a life that was imagined as simpler, slower, and closer to nature (Blickle, 2002; Clausen, 2008). *Heimat* is often called an anxious concept because it tends to be mobilized in conditions of insecurity and uncertainty, and this nostalgia is always for a place/time that to some extent never really existed, but rather is imagined and idealized (Blickle, 2002; Costadura et al., 2019).

Because it represents an emotional nexus of identity, belonging, and space, *Heimat* can be a highly problematic concept, despite also being a ubiquitous word in the German language (Blickle, 2002). During National Socialism, *Heimat* became synonymous with the nation, and was part of the concept of the ideology of blood, nation, and land (Räthzel, 1994). Many theorizations of the concept note that it has been instrumentalized for racist purposes, but they lack a race analysis. Some like Boa and Palfreyman (2000) assert that *Heimat* has no inherent racial content. Yet the subject of *Heimat*, as a concept born in the German-speaking world at the height of European colonialism, must be the modern, Western, apparently universal subject that is constituted through race (Silva, 2007). As a specifically *German* subject, it is doubly constituted in distinction to its racialized others (El-Tayeb, 2016). As a result, *Heimat* can only be seen as neutral if the whiteness of Germanness is ignored along with the colonial context in which it emerged.

Heimat has been front and centre in political debate in Germany recently, and Costadura et al. (2019) argue that the concept is currently having a general "renaissance." This is both evidenced by and a result of the announcement in 2018 of the addition of a new "Ministry of Heimat" (Heimatsministerium) to Germany's Interior Ministry. In the current context of anti-Muslim racism and anti-refugee discourse and policy in Germany some commentators – particularly racialized German scholars and journalists – interpreted the introduction of the ministry as an appeal to the racist right-wing for which Heimat often appears in narratives about defence and recovery. The AfD, for example, styles itself as *Heimatpartei* (Sanyal, 2019a). Interior Minister Horst Seehofer claimed that Heimat should be understood as referring particularly to rural Germany and not to a nationalist past, but he undermined his own claims by making a statement shortly thereafter that "Islam does not belong to Germany" ("Der Islam gehört nicht zu Deutschland"), and thus drawing a rather clear line regarding the subjects of German *Heimat* (Ataman, 2018a, 2018b; Aydemir & Yaghoobifarah, 2019).

A key to the debate about *Heimat* is whether one considers it an inherently exclusive concept or one which carries historical baggage that can be jettisoned to one's own purposes. Writing on the *Heimat* movement (*Heimatbewegung*) in Wilhelmsburg in the early 20th century, Sigrun Clausen (2008) emphasizes the historical nature of the *Heimat* concept, which emerged from particular values and interests and was cultivated by elite men. Clausen writes in part in explanation of the 2006 decision by the *Heimatkundeverein* to change its name and the name of its *Heimatmuseum*, which she reports involved rather intense and emotional debate within the association (Clausen, 2008). The *Heimat* movement in Wilhelmsburg was organized around historical preservation and museumization of the island's past in a time and place where industrialization and urbanization was rapid and intense (Clausen, 2008), as was the rise in immigration, as I discussed in the previous chapter. In what Clausen calls one of the paradoxes of the *Heimat* movement, the leaders of it had never lived the dairy farming life they wished to archive, but rather were city-dwelling men of the local elite, teachers and newspaper publishers who were likely disturbed by the rise of a new financial and industrial elite (Clausen, 2008). To Clausen it is thus no surprise that *Heimat* discourse is often patriarchal, turning on a construction of home and safety that is maintained by mothers and wives (see also Blickle, 2002; Boa & Palfreyman, 2000). *Heimat* is an imagined past that reflects particular people's desires for something they do not have (Räthzel, 1994).

An explicit race analysis of *Heimat* demands not only awareness of how the concept has been mobilized in Germany's (Nazi) past, but also that the concept has built-in troubles, because "it contains within itself its negative and other" (Boa & Palfreyman, 2000, p. 28). As a place with which one identifies, it depends upon a notion of "Other" people and places with which one does not identify and does not feel at home. In particular, it aligns with a definition of "Others" – migrants, foreigners – as antithetical to home (see Boccagni, 2016). Racialized German authors and scholars are thus taking multiple approaches to the concept's "renaissance," arguing variously for extreme caution with it (Ataman, 2018a, 2018b), its transformation into something that is explicitly plural and inclusive (Lamya Kaddor, quoted in Sanyal, 2019a), and its abandonment altogether to its right-wing, racist career (Aydemir & Yaghoobifarah, in Lichtblau, 2019). In the introduction to a volume of essays entitled *Your Heimat is our Nightmare (Eure Heimat ist unser Albtraum)* Fatma Aydemir and Hengameh Yaghoobifarah (2019) argue that *Heimat* must always be understood as connected to racist and fascist ideology, violence, and genocide. *Heimat* is "always a longing for an ideal, a homogenous, white, Christian society, in which men have the say, women mainly take care of having children, and other lived realities simply do not appear" (Aydemir & Yaghoobifarah, 2019, translation by author).[3] This reading aligns in many ways with feminist sociologist Nora Räthzel's (1994) earlier findings on how people in Germany and Switzerland conceptualize *Heimat* and *Ausländer*, which I have translated as "foreigner," but which she argues means much more literally "people who belong outside the country" (p. 86). Räthzel argues that "the term [*Heimat*] is only meant for those whom the dominant powers and 'common sense' understandings define as belonging to the nation. Those who do not belong can merely claim to feel at home but are not allowed to use the emphatic term Heimat that evokes the idea of roots, of having been there forever" (Räthzel, 1994, p. 89).

Beautiful, green, island village: What residents value about Wilhelmsburg

Wilhelmsburg has not, historically, been known as a green and beautiful neighbourhood. Instead, as the previous chapter described, the island has a long history of association with the dirty and loud functions of the city, and its relationship to the rest of the city has been defined by this association. Yet when I asked Wilhelmsburgers about the island and about what it was like to live there, residents' perceptions of the island as a green and beautiful place often took centre stage. The island's landscape emerged as a powerful anchor that connects residents to the neighbourhood. Bayram Inan (interview, 2018) told me for example that "Wilhelmsburg is actually a very beautiful, green island. That is wonderful. So that's why I always say I can't imagine living anywhere else." Inan has been a resident of Wilhelmsburg since he arrived as a so-called guest worker in 1973, the final year of the "guest worker" program. Today he is an engineer and a politician with *Die Grünen* who represents Wilhelmsburg at the district level.

Inan was not alone in feeling that anywhere else would pale in comparison to Wilhelmsburg's green. In fact, Arzu and her husband, both born and raised on the island, considered moving to a different central and in-demand part of the city, but thought better of it:

> It is incredibly green here; it is really great... At one point we planned to move to Hamburg, to Altona or Eppendorf or something, but then we looked at apartments there. We looked at the locations, and we didn't like it, we simply didn't like it, because it is just a more beautiful landscape here. And we found that what

3 In a disturbing development, in 2020 Heimatminister Seehofer announced – and then withdrew – a criminal complaint against journalist Hengameh Yaghoobifarah, arguing that they had written "unspeakable" things about the police in a column about abolition (Eydlin & AFP, 2020).

> we could get there, in terms of apartments and quality, we would have had to pay double or triple the amount. It just didn't make sense – because we thought, we get a much more beautiful neighbourhood, in terms of the space, right, and we get a better apartment for half the price – why in God's name would we move away?! (Arzu, interview, 2017)

Wilhelmsburg's beauty – and cost – won out, though as a double-income professional household the family could afford to move to a higher status neighbourhood. Arzu found value in the ability to go out one's front door and find a quiet, green spot within minutes, something that residents suggested is just is not the same in other parts of Hamburg. Mostafa, who has been a self-described "Islander" since 1981, felt similarly and also made a comparison to supposedly more desirable parts of the city.

> Mostafa: I would never trade Wilhelmsburg, not even for Blankenese [one of the richest neighbourhoods in Hamburg]. Because one is at home here, this is our home.
>
> Julie: What makes it home, do you think?
>
> Mostafa: Nature! The nature! … It is incredibly beautiful. My in-laws [who were recently visiting from Turkey] didn't want to leave. Early mornings, shortly before six, I'm already up and I go walking here. An hour, sometimes two hours. In Wilhelmsburg it is so gorgeous; one can't even express it with words.

As we talked, Mostafa led me along leafy paths to the lush growth around the Dove-Elbe, a small branch of the Elbe River. We passed small houses and gardens and some of the old farmhouses on the island. Mostafa used the landscape to demonstrate the beauty of the island and his affection for it, at every turn pointing out another delight.

Along with the island's green landscape the people of Wilhelmsburg themselves make the island special, according to the residents I interviewed. In stark contrast to how it has historically been portrayed in the media, they argued that Wilhelmsburg is a warm and familiar place. Gülhan, for example, has lived in Wilhelmsburg for over 25 years and works at a local social service organization. She was keenly aware of how Wilhelmsburgers have been racialized and criminalized in the media, as I noted in the previous chapter. What bothered her in particular about this, and about how the neighbourhood tended to be looked down upon from the other side of the Elbe, was that it was so contradictory to her own experiences. To Gülhan, Wilhelmsburg "is a very friendly neighbourhood, people are really welcoming. And yes, I think more should really have been built here, and the people upstairs should have paid attention to people's needs. We are really a cozy, sweet neighbourhood

Figure 13: A big, old tree on what appears to be an abandoned lot, Rotenhäuser Damm

(photo by author, 2019).

Figure 14: The Dove Elbe, Mostafa's favourite spot on the island

(photo by author, 2017)

Figure 15: The diverse forms of Wilhelmsburg: The rooves of sheds on allotment gardens peek through greenery, while a highrise overlooks

(photo by author, 2017)

Figure 16: Mare and foal on a farm near Finkenriek

(photo by author, 2017)

that really is not like is reported in the media" (Gülhan, interview, 2017). People on the other side of the river tended not to know what life on the island was actually like, from Gülhan's perspective; they viewed it from a distance and compared it with other neighbourhoods on the basis of what they thought was important. But Gülhan argued that "if you live here, you're happy to live here! You know yourself that it isn't like that and after a while you just don't listen anymore" (Gülhan, interview, 2017).

The "coziness" and "welcome" that Gülhan described was often attributed to the sense that Wilhelmsburg is a "village." Mohammed, who is a Wilhelmsburger born and raised and has a dental surgery in the neighbourhood, said for example that "Wilhelmsburg is a village with 50,000 people. Everyone knows each other here...[T]hat's why I simply feel at home here. Eppendorf[4] is ... I think there it is much more anonymous. It is much more 'what are you wearing, what am I wearing,' and that's not my thing." Not only is Wilhelmsburg not anonymous, in his view, it is also not superficial; people see each other and engage beyond the surface level. Paula, who has lived in Wilhelmsburg since the '70s, tends to agree. She told me that "to me this isn't the city, the city is anonymous" (Paula, interview, 2017). In Wilhelmsburg you can make friends on the bus, as she recently did, or go outside and find familiar faces when you are bored or having a bad day, as she would often do when her kids were young. People in Wilhelmsburg know each other, she argued, and as in a village, this lack of anonymity can be both a blessing and a curse. At times people notice what your kids are up to, or that you stayed home from work, when you perhaps wish they did not. Yet the feeling of being seen and connected was described in overwhelmingly positive terms: "I have definitely found a lot of nice neighbours. It's really nice... When I come home from work I might be invited to dinner, or 'come over, we're grilling!'" (Paula, interview, 2017). This sense of closeness in daily connection was often described to me as one of the best things about the neighbourhood.

Diversity is a key part of what makes the neighbourhood welcoming: a variety of different people live in Wilhelmsburg "more or less peacefully alongside one another" (Murat, interview, 2017). In contradiction to the dominant narrative, residents talked about this not as a problem, but as a part of the island's charm, noting its diversity in terms of "foreigners," "migrants" and "Germans," different language groups, Muslims and Christians, classes, and sexualities. Joseph told me for example that "there are a lot of foreigners here in the neighbourhood, and also there are Germans as well. So, we can easily mix up here than some other areas" (Joseph,

4 A part of Hamburg that is considered desirable to live in: central, with charming architecture, an average income almost three times that of Wilhelmsburg, and one-third as many residents "with migration background" (Statistisches Amt für Hamburg und Schleswig-Holstein, 2018) (see Figure 3).

interview, 2017).[5] Originally from Nigeria, Joseph leads a small church congregation on the island that is attended mainly by West African Wilhelmsburgers and their kids. After living in other parts of Hamburg for years, and now on the island for a decade, the mix of people in Wilhelmsburg is his favourite thing. Umut, who grew up in and around his family's convenience store in Reiherstieg over the past two decades, feels similarly. He uses a vivid metaphor to capture the intensity of diversity from his perspective:

> I describe Wilhelmsburg as the central station. When you arrive at the central station, you meet everybody. Regardless of social class, they all have to travel and arrive at their destination. Everyone meets at the central station, and at the moment I still have the feeling that Wilhelmsburg is that kind of place [*laughs*] (Umut, interview, 2017).

Wilhelmsburg's charm, in Umut's opinion, is that it is a meeting place for "everybody." People of various classes and backgrounds are present here, brushing up against one another, not necessarily sharing the same "destination," but nonetheless sharing the space.

Yet other residents described this sharing of space in closer terms. Whereas Umut's central station metaphor puts everyone in transit – in each other's presence, but not necessarily in contact – Yücel and Murat both stressed *familiarity* across difference as part of the character of Wilhelmsburg. Yücel, who has lived on the island for over 30 years, argued that "in Wilhelmsburg people know each other, as Muslims, Christians, Chinese, Eastern Europeans or whatever, we know each other" (Yücel, interview, 2017). In contrast to the reputation for conflict and problems, Yücel described a peaceful neighbourhood, in which people have demystified difference through the building of relationships over time. To make a similar point, Murat told me a story from his own life and encouraged me to reflect on my own perceptions of the neighbourhood:

> Murat: Here people know each other, people greet each other here. Of course, things here are very familial, as you'll notice too.
>
> Julie: Yeah, I notice it on the street and that. And now, since I've been here about three months, I notice that my neighbours – not only in the building – that my neighbours recognize me and say hi to me. It's really nice, you don't find that often where I live in Toronto.
>
> Murat: ...So at our place for example, our neighbours are homosexual. And they're so sweet – we often eat together, my mom invites them over, really nice guys. I

5 Quotes from Joseph have not been translated, as the interview took place in English.

> didn't realize at first. I asked: 'Mom, why are they always together?' and she said, 'I don't know, it doesn't matter.'

With this anecdote, Murat illustrated the everyday closeness of people who he perceives as different from himself. This is the familiarity of Wilhelmsburg, he argued, which is part of its beauty: neighbours with different sexualities are neighbours with whom one shares food, time, and one's home.

Murat's choice of example from his life pushed back against a racist trope in which racialized men like Murat and neighbourhoods like Wilhelmsburg are imagined to be homophobic and dangerous to LGBTQ people. This is tied to the production of the neighbourhood as foreign, Muslim, and violent regardless of how its residents self-identify (see Yildiz, 2009) and regardless of the daily reality of the neighbourhood, as Gülhan touched on above. As queer of colour scholars from Germany have argued, a key manner in which racialized people are evicted from Germanness in the context of contemporary racism, and anti-Muslim racism in particular, is through their imagined failure to embody values of tolerance and equality. In this trope, LGBTQ People of Colour are treated as either nonexistent or exotic (El-Tayeb, 2012; Haritaworn, 2015). As I discussed in Chapter 1, this contributes to the eviction of racialized people from urban spaces in the name of safety and security, and obscures homophobia and gender-based oppression that is perpetrated by the dominant (white German) population (El-Tayeb, 2012; Haritaworn, 2015; Petzen, 2009; Tsianos, 2013). Murat's description of his familiar relationship with his neighbours thus resists a common discourse about racialized neighbourhoods like Wilhelmsburg.

Residents also challenged other common stereotypes about racialized people and neighbourhoods, and particularly about Turkish speakers in Germany, when they highlighted the importance of Wilhelmsburg's linguistic diversity to their experiences of the neighbourhood. Mohammed noted for example:

> You get by in Wilhelmsburg without German. It's no problem: there are Turkish shops, Turkish restaurants, Turkish teachers, everything. Many say that's somehow a 'Turkish problem.' That's total nonsense. In the U.S. there's a Germantown, Little Italy, Chinatown. It's totally normal that people come together. If you met a bunch of Canadians here, you would do the same thing. Totally normal (Mohammed, interview, 2017).

Mohammed is referring here to how neighbourhoods in which Turkish speakers are concentrated tend to be pathologized, and Turkish speakers themselves imagined as failing or refusing to integrate, or indeed as rejecting Germanness (cf. Çağlar, 2001; Çalışkan, 2011; Ehrkamp, 2006). Mohammed argues that what tends to be seen as a problem is in fact normal. In the process he naturalizes the concentration of immigrants and racialized people in Wilhelmsburg, among them a significant portion of Turkish-speakers, which I demonstrated in the previous chapter is not in fact the result simply of people who have something in common "coming together." Neither were Chinatowns created in North America through residents' choice, but rather through policies and discourses that simultaneously produced and segregated "Chineseness" (Anderson, 1991). Yet by normalizing what he sees as enclaves (see also Hinze, 2013) Mohammed aims to unsettle the "parallel society" (*Parallelgesellschaft*) discourse which problematizes segregation in German cities as a matter of cultural and religious failure/refusal on the part of migrantized people. While this premise has been extensively debunked (cf. Gruner, 2010; Harlander, 2012; Münch, 2009), it remains an influential aspect of integration discourse and urban policymaking.

Indeed, the dominant discourse influenced how residents who – unlike Mohammed – are themselves Turkish speakers talked about the importance of the language to the warmth, welcome, and familiarity of the neighbourhood. An underlying assumption that it is bad to not learn German emerged from my interviews with three women who I met through a German class on the island, who had all immigrated from Turkey about 20 years earlier. Ayşe argued that the fact that Turkish is widely spoken is one of the neighbourhood's great strengths, but she also qualified that assessment:

> In Wilhelmsburg it is advantage and disadvantage: A Turkish shop, Turkish doctor, and Turkish – for example at Marktkauf or Budni [the grocery store or drugstore] there are Turkish staff. Turkish baker. It is a disadvantage, but if I don't speak German, advantage. Newcomer Turkish, or Turkish woman – doesn't matter, advantage, everything Turkish (Ayşe, interview, 2017).

The resources that Ayşe described in the neighbourhood were established over many years of immigration and of Turkish-speaking emplacement (Çağlar & Schiller, 2018), and made her life easier for decades, as she was able to meet her daily needs and the needs of her family. Yet those resources did not help her to learn German until recently, which she seemed to see as negative. Meryem, who lives in the Bahnhofsviertel with her husband and kids, offered a similar assessment. On the one hand, she likes about Wilhelmsburg that she has "lots of friends, neighbours...at the shopping centre I know everyone. Many Turkish people live here, and I visit them... I am always at home, and I go to my neighbour, neighbour comes to me. I think this is good" (Meryem, interview, 2017). Yet she also specified disapprovingly that she does not have relationships with "Germans" (which I will discuss in Chapter 6) and lamented that it had taken her decades to get to a German class. She seemed to see this both as negative and as her own fault. Zehra, who I interviewed together with Meryem, pushed back against this self-deprecation and claim of failure, pointing out that Meryem had been raising kids and that that was more than enough work at one time. Put together with

Ayşe's analysis of advantage and disadvantage, the exchange illustrated the idea that one *should* urgently learn German after immigrating, an idea that is closely tied to the integration discourse I introduced in Chapter 1. According to prevailing myths of integration (cf. Kaddor, 2010), immigrants should always be striving to integrate themselves, particularly through education, in order to be seen as willing (K. N. Ha, 2007a), as loyal to Germany (Çağlar, 2001), and as potentially useful to the nation (see Schiffer, 2010). On the other hand, Zehra demonstrated an analysis that gendered responsibilities for children impact how and whether one can do this striving. Despite the resulting ambivalence in the discussion, the Turkish-speaking community in Wilhelmsburg was at the centre of the women's attachment to the neighbourhood, as a practical resource and a key to crucial social supports.

Wilhelmsburg's long immigration history has also produced a neighbourhood in which racialized people and immigrants see themselves reflected. This is the case in terms of language, but also in the general visibility of racialized people living out their lives, which was described to me as something special about Wilhelmsburg. Paula, who has lived on the island since 1976, argued, for example, that this distinguishes Wilhelmsburg from other parts of the city: "It is just like that in Wilhelmsburg. It's just normal here to see whatever kind of immigrant at the cash, wearing a headscarf or whatever. I don't know if it is like that in all neighbourhoods, here everything is accepted" (Paula, interview, 2017). Nalan made a similar observation, and described it as important to her:

> I know this also from many others who moved to another neighbourhood, for example, where there aren't so many migrants, where it is not so multicultural. They don't feel comfortable there. They're looked at funny, like 'What are you doing here?' And you just don't have that feeling here, when you're out. It is totally normal, at the pharmacy you also see many, many who for example work with the headscarf, you don't see that anywhere else (Nalan, interview, 2017).

Both Paula and Nalan use the word "normal" to describe the visibility of "migrants" in all aspects of life in the neighbourhood. To Paula it is a sign of acceptance, and of how racialized people have shaped the space over time. For Nalan the significance is that one blends in, does not receive unwanted attention, and is not told through looks that one does not belong. For both there is a crucial counterpoint against which Wilhelmsburg is contrasted: other spaces in which racialized people *do* stand out, *are* viewed suspiciously, and are *not* accepted.

Their choice of examples reveals the important role that women's bodies, and here particularly Muslim women's bodies, play in defining the boundaries of belonging (Anthias & Yuval-Davis, 1989; Mohanram, 1999; Yuval-Davis, 1997). Both residents mentioned women in headscarves working in customer-facing jobs as a symbol of the normalcy and visibility of migrants. It is often the case in the Ger-

Figure 17: The Spreewald seen here is a small forest that developed on land abandoned by the city after the 1962 flood

(photo by author, 2017)

Figure 18: In Wilhelmsburg the green often butts up against industry and container yards, as it does here at Reiherstiegsknie

(photo by author, 2018)

Figure 19: One of many places one could expect to find Turkish spoken in the Bahnhofsviertel

(photo by author, 2017)

Figure 20: Trees line Vogelhüttendeich in Old Wilhelmsburg, where one can find Balkan groceries and a fresh mural

(photo by author, 2017)

man context that the "woman in headscarf" is a shorthand for "Muslim," "Turkish" and "immigrant," and indeed for so-called problem neighbourhoods (Ramm, 2010). However, Paula and Nalan use the figure as a shorthand for racialized belonging and as a specific example of belonging in the context of exclusion. In so doing they implicitly speak back against the quality of the dominant discourse. They are not simply reproducing the typical symbolism, but turning it around; for Nalan in particular, as a woman who herself wears a headscarf, the symbolism is personally meaningful.

On the island, Muslim women who wear headscarves are visible in positions from which they are elsewhere often excluded. Muslim women's head and body coverings have been a focus of European fascination and fear for centuries (Yeğenoğlu, 1998), and since 9/11 have particularly been used as an overloaded symbol to illustrate the supposed superiority of white, German Christianity (Yildiz, 2009). In Hamburg and in Germany it is common for employers not to allow hijabs (Islamic head coverings) in the workplace, though Muslim women have mounted legal challenges to this exclusion (cf. Leoni, 2018). As I am writing there is no clear legal standard in Hamburg, as the federal court has allowed each state to decide its own stance on the issue, and a question was recently sent to the European Court for an opinion on whether forbidding hijabs is consistent with European law (Fuhlrott, 2019; Hartmann et al., 2019).

In this context, discrimination against Muslim women and girls who wear headscarves is one of the biggest challenges facing young women Wilhelmsburgers, according to Mädchentreff, an organization for girls in the Kirchdorf-Süd part of Wilhelmsburg (personal communication, 2017). Several residents also told me about specific experiences of discrimination in education and employment. In one instance, a bank worker who decided during maternity leave to start wearing a headscarf was told when she returned that she had to choose between the job and the headscarf (personal communication, 2019). In another instance, the offer of a practicum position for a teacher trainee was made conditional upon removing her headscarf (Mohammed, interview, 2017). Nalan herself described in detail one of her own experiences when she went to a job interview with "a large, well-known institution":

> I was invited for an interview, and there it was said, 'yes, I find, I have no problem with it [her headscarf], but the bosses, those at the top, they are very conservative. They don't want that. That would be a problem.' 'OK, why did you invite me here then? To tell me that?' ... Some don't say it to your face, but lots just think like that. He said it to my face. Good to know... A lot of migrants face a hurdle anyways, and then you add to that that you're a woman. That's hard too. And then add to that that you wear a headscarf. So, three hurdles at once (Nalan, interview, 2017).

In each of these women's stories, the discrimination was framed as a "choice" for the woman to make, and in every case the woman chose in favour of her religious observance as she understood it, and against working for an employer who had revealed anti-Muslim practices.

These experiences illustrate the potency of the symbolism of women in headscarves in Wilhelmsburgers' own accounts of the neighbourhood, as they demonstrate some of what Muslim women are up against not only in the job market but in society more broadly. In the context of Muslim people's struggle for belonging in Germany, a headscarf is more than just a headscarf: it is the focus of anxiety about German identity and society (Yildiz, 2009) and a key to the narrative about dangerous Muslim men from which Muslim women must be protected (or, as Sherene Razack (2008) succinctly calls it, "racism in the name of feminism"). To the extent that women are constructed as reproducers of the nation both biologically and in terms of transmission of culture and tradition, they also often appear as "symbolic border guards" that define the boundaries of a collectivity (Yuval-Davis, 1997, p. 23). The behaviour of women, how we dress, and what we do with our bodies is a crucial field of symbolism and control through which collectivities distinguish and define themselves (Anthias & Yuval-Davis, 1989). The residents I spoke to in Wilhelmsburg saw women wearing headscarves not only as normal and acceptable but as symbolic of the boundaries of belonging in the community, which are different than the boundaries of the dominant society. Crucially, the normalcy and visibility of Muslim women in headscarves as described by Paula and Nalan is partial and incomplete: the bank and school for which headscarves were unacceptable, for example, are both in Wilhelmsburg. Yet the neighbourhood remains remarkable for having any space at all for visibility and acceptance.

This is also the case when it comes to the presence and condition of Islamic religious spaces on the island, from Wilhelmsburgers' perspectives. The presence of multiple mosques, of a Muslim cemetery, and soon of a wash house in the cemetery, is a factor that facilitates a sense of belonging particularly for some of the Muslim residents I spoke to. Özden, who has lived in Wilhelmsburg for 20 years, told me for example how much he appreciates that in Wilhelmsburg there are mosques in close proximity to his home and work. "It is important to me. When it is time to pray, I want to have a mosque in sight" (Özden, interview, 2017). That there are a variety of different congregations nearby contributes to his sense of feeling at home and accepted in Wilhelmsburg, because he can meet his religious obligations without having to leave the neighbourhood.

Yet the location and condition of Muslim prayer spaces is a sign of Muslims' ongoing struggle for dignified space within Germany and Hamburg in general. Many of Wilhelmsburg's mosques are in basements and backyards, or in small rooms that do not fit their congregations. Several residents thus argued that given the number of Muslims in Wilhelmsburg – a number that is not precisely known, but that has been estimated at 17,000 to 18,000 people[6] – the neighbourhood is short on what Mostafa (interview, 2017) called "decent prayer space." Bayram Inan (interview, 2018) offered an example: "How the mosque in Kirchdorf is now, it is pretty small, the landlord gave a small room, the Friday prayers are in the courtyard, in the garden, sometimes in the rain."[7] Inan's hope for many years was that the different mosque communities would come together to support the building of a large, central mosque in Wilhelmsburg "with a beautiful minaret, even" (Inan, interview, 2018). The idea got as far as discussions and identification of a location in the midst of the recent redevelopments in the neighbourhood (which the next chapter will describe). A possible plan was rendered, but the project stalled, and residents I spoke to despaired of the likelihood that it would be revived (see also Bezirksamt Hamburg-Mitte, 2015; Hamburger Abendblatt, 2016).

A similar state of affairs is also reflected in Hamburg more broadly, though this is perhaps slowly changing (see Haddad, 2018), and in other cities in Germany. Mosques often "find themselves in materially hidden spaces" in German cities, and communities' efforts to change this by renting, buying, or renovating are often precarious (Kuppinger, 2011). Muslim prayer space is both a practical and symbolic issue of visibility in urban space, and struggles to achieve it are fundamentally political, as they are about the position of Muslims in German society as much as they are about the physical space (Kuppinger, 2011, 2014). Several Wilhelmsburgers themselves argued that the slow-moving struggle for dignified Muslim prayer space was shaped by Islamophobia and by anti-Muslim racism in Germany and in Hamburg. Mostafa felt for example that the idea of a central Wilhelmsburg mosque had been set up to fail. On the surface there was support from the city, but the support was for a project that he felt was clearly unattainable and unsustainable (Mostafa, interview, 2017). Mohammed described his perceptions of widespread hostility towards Muslims in Germany and in Hamburg, which shaped the struggle for space. As he pushed an imaginary button on the table of the café where we were sitting, he said: "If there were a button that said 'all Muslims out,' they would push it" (Mohammed, interview, 2017).

In this hostile context, the building of a wash house in the Muslim section of the cemetery at Finkenriek in the south of Wilhelmsburg is an important recent

---

6 This estimation comes from Bayram Inan (Die Grünen), representative in the regional council, but it is not clear how the estimate was achieved (Trautwein, 2014b). No statistics are gathered on religious affiliation. A 2010 report estimates 90,000 Muslims in Hamburg, with 71,000 living in Hamburg-Mitte, the district in which Wilhelmsburg is located (Open Society Institute, 2010). More recently the Heinrich Böll Foundation estimated that 150,000 Hamburgers are Muslim, but also without explanation of how they arrived at the number (Hamburger Abendblatt, 2016).

7 A scene in Schaefer's (2013a) video ethnography of the Wilhelmsburg bus route, Die Wilde 13, depicts this situation.

achievement for the visibility and acceptance of Muslims in Wilhelmsburg. Since 2010 local Muslim activists and politicians, including Mostafa and Bayram, advocated for the construction of a prayer- and wash house in the cemetery, so that the rituals required for Islamic burial could be fulfilled on the island. Previously, bodies had to be transported across town to the one other Muslim cemetery in Hamburg to be prepared for burial, which took a financial and emotional toll on families (Trautwein, 2014b). This advocacy has been successful, and in May 2019 the first stone was laid for the foundation (Deutsche Einheit Fernstraßenplanungs- und -bau GmbH, n.d.).

Yet it was a long and discouraging road for those involved, and ultimately came as a result of a highly controversial highway development on the island. DEGES, the corporation responsible for the development of transportation infrastructure, supported demands for the wash house in exchange for cooperation with the reinterment of bodies to make way for the highway. I thus found it both celebrated as an achievement for the community and lamented as an example of the disinterest of the various levels of Hamburg government with the needs of Muslims, and of immigrants and racialized people in general. For almost a decade, community advocates had experienced a pattern of presenting the case and making a recommendation to the local council, only to have the district administration refuse to implement their recommendation (Trautwein, 2014b).

On the one hand, this is characteristic of the hierarchy of local government and the governance of development in Hamburg: the insitutions that are closest to the everyday life of the neighbourhood can make recommendations but have the least power to implement decisions. Yet advocates felt that there were other forces at play, and suspected that the city-state wanted to avoid any protest from other city residents, which is common when it comes to the building of mosques. This would be consistent with the lack of support that municipalities routinely display in relation to processes of Muslim place-making (Kuppinger, 2011, 2014). My sense is that residents' perceptions are not always taken seriously by people in positions of political power. When I, for example, suggested in my interview with the Chairperson of the Wilhelmsburg Development Council, that Muslims are struggling for belonging in Germany in the context of anti-Muslim racism, his response was to ask: "What do most of the Muslims who live here have in common with the AfD? They play the victim role" (interview, 2018). Certainly, some of the residents I interviewed complained of a lack of political will and sense of urgency when it came to issues that Wilhelmsburgers find pressing. I found this a troubling way to avoid the discussion as it seemed both to trivialize the AfD's politics *and* residents' real concerns about marginalization.

These threads bring home the reality that racialized people in Wilhelmsburg face exclusions, discriminations and structural oppressions that span far beyond Wilhelmsburg, and this is the context in which attachments to Wilhelmsburg are

Figure 21: Diyanet İşleri Türk-İslam Birliği mosque during Ramadan, with outside space set up for communal fast-breaking

(photo by author, 2019)

Figure 22 (left): Said-i-Nursi, a storefront mosque at the foot of Veringstrasse; Figure 23 (right): Until recently you had to look carefully to find the sign for the Ayasofia mosque where it was located in a back courtyard on Vogelhüttendeich. Its new location, seen here without a sign but with a banner for Ramadan on its windows, is larger and more visible

(photos by author, 2019)

Figure 24: The Muslim section of the cemetery at Finkenriek

(photo by author, 2017)

formed. The discriminations are based in the very process of racialization, which evicts some people from Germanness and thus attempts to exclude them from being at home in the nation. This has material consequences that arose from time to time in my interviews, in stories of discrimination in schools and marginalization from work opportunities on the basis of "race." In the following section I discuss some of the attempted exclusions– *attempted* because Wilhelmsburgers often described practices of resistance in the same breath – before going on to explore the significance, in this context, of Wilhelmsburg as *Heimat*. *Heimat,* while a problematic and indeed exclusive concept itself, emerges as a critical claiming of Wilhelmsburg as home, as the place where residents feel comfortable, in a broader context that often makes them uncomfortable and excludes them from belonging.

Migration foreground and struggles for belonging

"If I ate pork, drank alcohol, and went dancing with my wife, it would be 'yes, you are a good foreigner.' Not one of them: I am a good foreigner." This quote from my interview with Özden (interview, 2017) illustrates succinctly what many of the racialized Wilhelmsburgers I interviewed brought up in various ways: there are persistent structures and attitudes in Germany that keep certain people – Muslims, migrants, and people perceived as either – on the outside of Germanness, even though they are a part of German society. These are part and parcel of racialization as Fatima El-Tayeb defines it: "the attribution of collective quasi-biological and or/cultural qualities that allow the perception of certain groups as not belonging, even when they are already part of society" (El-Tayeb, 2016, p. 34, translation by author). The qualities appear to be immutable; as Özden says, and other Muslim Germans such as Lamya Kaddor (2010) have argued elsewhere, he can abandon practices that he considers to be part of his Muslimness, but he will still be considered "other" anyway. The possibility of somehow integrating into Germanness is a myth because Germanness is constituted as a racial category, in constant contrast to "Others" (El-Tayeb, 2016).

The most that one can do is try to blend in, which Paula describes having done over time in response to painful experiences of exclusion as a "foreign" kid in the 1970s. "My kids always say I'm more German than the Germans, and that might just be because I was so affected by it that I thought 'no, I don't want them to think that I am different, I'm just as good as all the others [*laughs*]. I also – when I got married, I took my husband's name, because I thought this long name always gives me away immediately" (Paula, interview, 2017). She speculated that she was able to blend in a bit, to pass for German at times, because her skin and hair colour did not particularly stand out, and she was not Muslim, and so was not targeted on that basis.

Wilhelmsburgers' experiences illustrate how the categories of "Other" change and shift while maintaining the central exclusion from Germanness. Özden's experiences point to the categories "foreigner," and "Muslim;" he is resigned to always being Muslim, but he wondered how many more decades in the country it would take to no longer be a foreigner (Özden, interview, 2017). From what I heard from other Wilhelmsburgers, the prospects do not look good. Some described being called "guests" long after the end of the so-called guest worker program, which I described in the previous chapter. "A guest stays maximum three days," as Bayram Inan (interview, 2018) put it to me 45 years after his arrival in Wilhelmsburg. In the 1970s, Inan worked as an engine fitter until he was able to pay his way through school for engineering; today he is a member of the local district council for Wilhelmsburg (for *Die Grünen*). He argued that the "guest" narrative continues in particular to affect members of the Turkish-speaking community, to which he is closest. The exclusionary discourse over many years has cultivated what he described as a sense that "we're just bystanders" in the country. This is something he tries to challenge in his political work. "You can't do that," he says, "you can't give people the feeling that 'they don't listen to us... I'm just reserve here anyway'" (Inan, interview, 2018). "Reserve" here refers to the logic of availability and disposability of migrant and "guest" workers that I described in the previous chapter. When people feel this way, Inan argues, they do not see themselves as the subjects of urban development processes that are supposed to be participatory and consult with the community.

For other Wilhelmsburgers, calling people "guests" is not only inaccurate but mobilizes the dehumanizing logic of the "guest worker" program itself, including the very idea that people who came to Germany to work would or should not put down roots. Mohammed positively spit out the word "GUEST workers! Take this word: 'guest worker,' so that everyone knows that they are only coming for a short time and then leaving... As though it is trade in goods" (Mohammed, interview, 2017). Özden (interview, 2017) argued that people were treated like livestock, valued based on their physical strength alone. His grandfather and great uncle came to Germany as part of the program, and quickly returned to Turkey after seeing a co-worker lose an arm in the workplace. Özden's encounters with the "guest" narrative are viewed through an analysis of apparent disposability and dehumanization. The damage that the discourse has done to Wilhelmsburgers' sense of belonging in Germany is particularly evident in Mohammed's account. Mohammed was born in Hamburg to parents who were refugees from Palestine via Jordan, yet as a person with dark hair and the name Mohammed, his supposed position was made very clear to him.

> Mohammed: At 10, 15, 20 years old I still often heard that we're guests here. Yes. I am supposedly a guest here.
>
> Julie: And what was that like?
>
> Mohammed: Disgusting. Disgusting. We grew up with it, we do not know anything other than that we are foreigners and we are different from the Germans, because we have always been told that. 'That's how you foreigners are.' (Mohammed, interview, 2017)

By citizenship and by self-identification, Mohammed *is* German, but these experiences deeply affected his sense of belonging in Germany. In contrast to residents like Özden, who immigrated to Germany himself 20 years ago, and thus said that he expects to some extent to be treated like "a foreigner" because, technically, he is, Mohammed is at home, has always been at home, and yet called a guest and a foreigner, illustrating the fundamentally racial nature of the exclusion.

The terminology has shifted to some extent over time, but the racialized exclusion from Germanness continues, including through the concept of "migration background," which as I discussed in Chapter 2, I used myself in this research as it is a racializing category used by the state and in common parlance. Murat argued that in the very use of the concept "there is actually already an exclusion" (Murat, interview, 2017), and drew attention to how it is taken up in everyday use. "I always call myself, when there are people who ask, 'German with migration background?' I say: 'No, migration foreground.' Because that's actually the whole joke: *they* associate me with it, *they* associate me with migration" (Murat, interview, 2017). Murat pushes back against that association, as someone born and raised in Hamburg, because he does not accept the logic that immigration defines who he is because it was something that his parents experienced. This association takes place regardless of how Murat associates *himself* with Germanness. Metin Hakverdi (SPD), who sits in the Bundestag (national parliament) for Wilhelmsburg and previously represented the island in Hamburg parliament, shared a similar analysis with me. "I will be 49 this year, and I have lived on the same street for 49 years. I'm not an immigrant. I am the opposite of an immigrant [*laughs*]... With me one can clearly establish the difference between 'migration background' and 'migration foreground,' and that migration background has almost no value to describe who someone is" (Hakverdi, interview, 2018). Hakverdi argued that he knew a lot about immigration from living in Wilhelmsburg, but rejected the implication that where his parents were born could automatically infer anything about him personally.

Imposition of the concept from the outside, which is the essence of how "migration background" functions, including in the census (Ahyoud et al., 2018), can challenge ones' own articulation of belonging. Murat went on to tell me a story about his first encounter with the term "*biodeutsch*," which in its common use is something like the opposite of German with migration background. Not a concept in Germany's official statistical lexicon, but rather something that one hears used informally, "*biodeutsch*" roughly translates as organically or biologically German, and has recently become a popular way to signify "white German" or "ethnic German," at times ironically, without having to say those words directly. When Murat first heard the term "*biodeutsch*" used at a conference, however, he understood it as referring to people with deep roots and attachment to Germany, and thus felt it applied to him. When he was asked as a panel presenter how he self-identified he said: "Well, I can tell you: I am married to this country. I feel at home here, and that is without a doubt. I'm probably '*biodeutsch*' now, I don't know.' And of course, everyone laughed" (Murat, interview, 2017). Everyone laughed because "*biodeutsch*" is a racial category that does not – cannot – include people with "migration foregrounds." Instead, people like Murat and many other Wilhelmsburgers must always be prepared to explain and defend who they are and where they belong in relation to the German nation (cf. Çağlar, 2001; Kaddor, 2010). Murat marvelled at the constant creation of new terms with which to marginalize people. "What are we supposed to do with that?" he asked me, a rhetorical question that he was not sure how to answer himself.

Another common manifestation of this demand to explain and defend is in the question "where are you from?". Mohammed (interview, 2017) described how often he is asked this question based on how he looks, and how the conversations tend to play out.

> Mohammed: It's like 'yeah, where are you from?' 'From Hamburg,' I always say. 'Yeah, yeah I mean originally.' I say: 'yes, from Hamburg, I was born here.' 'Yeah, hmmm, I mean your parents.' 'Oh, that's what you mean.' 'Do you want to go back?'
>
> Julie: Is that a question!?
>
> Mohammed: Yeah, very often! Where am I supposed to go back to? 'Yeah, to your *Heimat*.' 'This is my *Heimat*, I was born here. What do you mean by that, where should I go back to?' Then they start to stutter and realize what rubbish they're actually saying, and then sometimes I say, 'Don't worry, I'll be gone soon.'

This question of "where are you from?" is so commonly posed to racialized people that it has been described as a cliché (Sanyal, 2019a). Mohammed is intent on resisting the question and forcing his interlocutors to reflect on the implications of what they say, but that resistance has an energy cost. The cost is reflected in his quipping "don't worry, I'll be gone soon:" he has considered emigrating, going somewhere where he is in fact a foreigner, in order to escape the exclusion at the heart of the assumption that he is not at home in Germany.

The various terms of exclusion that Wilhelmsburgers identified in interviews illustrate the "differential racialization" that Laura Pulido (2006) has conceptualized. Racialized people are not homogeneous and do not necessarily share precisely the same experiences. Rather how groups are racialized is interconnected with the specifics of migration history, place of birth, class, religion, and so on, as well as with the particular political, economic and social historical moment (Pulido, 2006). All of the long-time residents I interviewed described distinct experiences with racialization that depended on where they were born, whether and when they came to Germany, their religion, and their class location. Yet while the specific history and terms of exclusion varied, the positioning as "other" from Germany was consistent.

These exclusions have practical and material manifestations in the lives of Wilhelmsburgers who are labelled as guests, foreigners, people with migration backgrounds. A most common manifestation is in structural inequity in schools, which is a well-documented issue in Germany (Davoli & Entorf, 2018; OECD, 2018), and one that came up in many of my interviews, second only to the problem of discrimination in housing, which I will discuss in the following chapters. Some Wilhelmsburg residents who had gone through the German school system described experiences of being streamed into non-academic school paths or pressured towards such streaming. As adults today they find themselves advocating for children who continue to experience this streaming, and encountering teachers who evaluate their performance and behaviour unfairly based on their perceived background.

This is consistent with the repeated findings of United Nations' reports on human rights in the country, which demonstrate that the German education system structurally disadvantages children who are racialized, immigrant, and poor, particularly through its streaming structure (UN Human Rights Council, 2007, 2010, 2017). The structure is such that young people are recommended by their teachers at quite a young age for streaming into academic or non-academic and trade-focused schooling. This can have lifelong effects, as it limits or opens up the range of options for higher education and shapes the kinds of work one can go into.

While this streaming system reproduces class stratification in general (Karakayali & Nieden, 2013) it is also a place where the reproduction of the racialized system of labour within racial capitalism is made visible. The streaming of racialized students into non-academic paths reproduces the racialization of low-paid work and unemployment, and in turn contributes to the labelling of racialized people as "*bildungsfern,*" (literally: distant from education). I first heard this term used in relation to Wilhelmsburg (cf. IBA Hamburg, 2010, p. 35) but it is also applied to other racialized neighbourhoods, often as a stand-in for race (Karakayali & Nieden, 2013). In the context of an educational system that is racist and classist, it falls to adults outside the education system to support, defend, and advocate for children to make their way through it. If they do not or cannot do this for any reason, it is considered a family weakness, rather than a failure of the education system.

The sorting of children and importance of parental advocacy begins as early as preschool in Wilhelmsburg. Arzu found for example that her daughter was grouped into an all "migrant" group of kids, in contrast to another group that overwhelmingly comprised "German" kids. As I will discuss in the following chapters in the context of claims and hopes related to "mixing," this problem reflects the power of (white) German parents and ongoing assumptions that "migrants" are the weaker students, an assumption that is partly based in prejudice, and partly in the history of precisely such practices of discrimination (Karakayali & Nieden, 2013).

Such experiences in Germany's education system then feed into experiences on the job market. Along with the exclusion from workplaces of Muslim women who wear headscarves, and the challenges of applying for jobs with a "migrant-sounding name" (Gizem, Mohammed, interviews, 2017), I also heard from Peter about a long-term struggle of African immigrants for decent jobs, for example. From his perspective as a pastor he observed a pattern of people coming to Germany and getting stuck in low paying and low status jobs for which they were over-qualified. Peter argued that to some extent the struggle for decent work is normalized as the immigrant journey – one must learn German and climb the ladder and so on. This is a narrative that justifies the exploitation of Bulgarian migrants, for example, who have arrived in Wilhelmsburg in recent decades and are experiencing precarity and mistreatment that I touched on in the previous chapter (see also Lintschnig & Schultz, 2018). What concerns Peter is that the pattern seems to persist over time.

> And some of us have been here for about 30 years. And these people, some of them, were lecturers in universities in Ghana. Some of them were university students, some of them were engineers. So, is that all that the German government can offer them? What are they doing to make these people useful, in their hands? So those are the challenges our people face (Peter, interview, 2017)[8].

As a result of these challenges, which he framed as a devaluation of the capacities of Africans, Peter discourages people who are considering moving to Germany from making that move. Özden told me how he himself struggled with what he experienced as a demotion and a major change in class position when he arrived in the country. "The first time that I wore work clothes in my life was here in Germany, working at the airport. It was – imagine, what kind of feeling that was. In Turkey there were five people under me, they worked for me. Because it was Germany I was supposed to start at the very bottom. I said, well OK, 'welcome to Germany!'" (Özden, interview, 2017). From Peter's perspective, however, even the second generation struggles more in Germany than they do in the UK. "This is not a place that you stay and you have a future," he argues (Peter, interview, 2017). Yet he has settled in Wilhelmsburg himself after years of migrating back and forth from the UK as the church congregation became established. "I am here because I am working for God, [otherwise] there is no way I would stay in this country [*laughs*]. There is no way" (Peter, interview, 2017).

Ali, who is a former resident of Wilhelmsburg and a counsellor at an organization serving immigrants, argued for a systemic analysis of a whole legal and regulatory regime in Germany that is hostile to the well-being of workers, of migrants, and of people who have lived in the country for years but are not German citizens. Over the past 20+ years of neoliberalization in Germany the stability of work and protection of workers has gotten worse. The minimum wage is low, argues Ali (interview, 2017) and workers are at the mercy of employers who are emboldened by the fact workers have short contracts and temporary status. All low-income Wilhelmsburgers struggle with this, however residents who do not have German citizenship are multiply vulnerable, as residence depends upon the working relations whose precarity is normalized. To this picture Ali adds that despite 40 years of migrant organizing around the issue, non-citizens continue to have no voting rights. Thus, there is a system of differential rights and entitlements that is not recognized as unjust because it affects people who are not considered to be German though they may, like him, have lived and worked in the country for decades. "What is that if not racism?" Ali asked (interview, 2017).

8 Quotes from Peter have not been translated, as the interview took place in English.

Romani residents of Hamburg are intimately familiar with the nexus of differential rights and state instruments that can be downright hostile. A local Romani organizer, who I met in 2017 as she was establishing a weekly café as a meeting point for Romanies on the island, taught me a great deal about the struggles of Romani people for the right to stay in Germany. Having fled the Balkans, she has lived in Germany for 25 years. Like many Romani people in the country, she had a *Duldung*, an exceptional leave to stay, which meant she would not be deported, but was also prohibited from working or studying. For 25 years this has held her and her family suspended between rights and rightlessness, as El-Tayeb (2011) puts it. Every six months she had to go to the Foreigners Authority to have the *Duldung* renewed. There she was encouraged to "self-deport;" at the time I met her, staff had escalated pressure on her by requiring her to return to renew her *Duldung* every three months. Far from an isolated case, many Romani and many refugees from civil wars live with *Duldung*, a practice through which the state attempts "to prevent the unavoidable, that is, people claiming the space they live in as 'home'" (El-Tayeb, 2011, p. 26).

Figure 25: Anti-Duldung graffiti on a renovated façade on Vogelhüttendeich, Wilhelmsburg

(photo by author, 2017)

Viewed within the framework of racial capitalism the differential rights and differential access to education and work that residents described clearly contribute to the reproduction of a racialized system of labour. Racialized people are sorted into different kinds of work by differential access to opportunity that manifests in a range of ways, from streaming in education, to exclusion from specific positions, to *Duldung*. And yet, in the resulting context of struggle for belonging, well-being, and indeed survival in Germany in the face of intersecting racialization, systemic racism, immigration status, and class, there are Wilhelmsburgers who expressed a deep sense of belonging and attachment to one place in Germany: to Wilhelmsburg. Often this sense of belonging and attachment was framed specifically through the term *Heimat*, which is a spatial concept that is tied to identity and belonging, as I defined in the introduction to this chapter (Blickle, 2002; Boa & Palfreyman, 2000; Eigler, 2014; Eigler & Kugele, 2012). I find residents' use of this particular concept significant, as I will discuss in the following section, as it is a contested concept marked by past and present uses in service of racism and nationalism. For Wilhelmsburg residents, to call the island *Heimat* was a way of expressing the feelings of attachment to the neighbourhood as a result of the qualities that they appreciate about it – the qualities of warmth, welcome, diversity, familiarity, and the reflection of racialized people, immigrants, and Muslims in daily life and space, as I have laid out above. Nalan for example made an impassioned statement.

> Nalan: You feel very comfortable here. I love Wilhelmsburg, really. I don't want to live anywhere else. We have the possibility, we could buy somewhere else, and this and that. We could live anywhere in Hamburg, but I like it here. It has something *heimatlich*, you feel comfortable here, here you feel like you have arrived. It is simply beautiful, to have such a thing, that you feel comfortable somewhere – second *Heimat*, or first *Heimat*. *Heimat*.
>
> Julie: That through that you are somehow –
>
> Nalan: Anchored. But you can hear that from almost everyone, that they have these feelings about it, *heimatlich* feelings. You can definitely say that. It is actually a strength.

I certainly did hear this from almost everyone, as Nalan suggested. The repetition of the theme of *Heimat* was striking in my interviews and others similarly argued that it was a strength in a neighbourhood that was not afforded a reputation for many strengths. Indeed, the feelings of *Heimat* just illustrate how little non-Wilhelmsburgers really understand about the neighbourhood. Gülhan argued that "when one lives here, because one lives here, one wants to stay here, because you are *heimisch* here," and she connected this feeling of being "*heimisch*" to a sense of security in contrast to other neighbourhoods. "When you are in Wilhelmsburg you feel very safe and secure. Maybe that's what you know. But the others who do not live here probably think 'oh, not a nice neighborhood,' but that's not the case. You just have to experience that people are friendly, nice. Different nationalities" (Gülhan, interview, 2017). The safety and security that Gülhan feels in Wilhelmsburg contrasts with its bad reputation, however Mohammed theorized that the two may in fact be connected. He argued that the bond that Wilhelmsburgers have with each other and thus with the neighbourhood, was forged not in spite of years of stigma, but because of it.

> Wilhelmsburg always had a very, very bad reputation in the rest of Hamburg. That's why Wilhelmsburgers are perhaps fonder of each other than the people in other neighbourhoods. Wilhelmsburgers know each other, everyone shares their neighbourhood. Everyone is always – somehow if I see a Wilhelmsburger somewhere, you're immediately good, you know: 'he's a Wilhelmsburger.' (Mohammed, interview, 2017)

The sense of being "immediately good," which I see as a sense of familiarity and trust, is a key ingredient for the *Heimat* that Wilhelmsburgers describe. Mostafa has lived in Wilhelmsburg for decades now, and said "it has become *Heimat* by now, after thirty-five years. You know everyone, you know people," he said (Mostafa, interview, 2017). Casim, who has also lived in Wilhelmsburg for decades described how he has come to "know everyone" through the roles he has taken on in the neighbourhood. He and I know each other from Haus der Jugend Wilhelmsburg, a recreational centre for kids and youth where I did a short internship on my first visit to the island in 2011.

> This is like our Heimat, we know everyone, we know the environment well. When we go out, everywhere you see an acquaintance, a friend, a colleague. I was involved with football for a long time, and through that I know a lot of families, and here in Haus der Jugend, and through that I know a lot of parents, their kids come here, and now their grandchildren. And that's why we feel comfortable here. Elsewhere we would feel foreign (Casim, interview, 2017).

The familiar relationships with which residents characterize Wilhelmsburg form the basis of it as *Heimat.* At the same time, it is again contrasted with other places where one would not feel so comfortable or at home because those relationships are not present. "Foreign" could also mean the kind of attention that "migrants" experience in other parts of the city, and the question of "what are *you* doing here?", a question that racialized residents do not expect to be asked in Wilhelmsburg.

In the following final section, I reflect on the picture that Wilhelmsburgers painted for me of the island, their *Heimat* that has a host of valuable qualities and strengths. I argue that Wilhelmsburgers communicate a sense of belonging, a subjective feeling of being at home (Castañeda, 2018) in Wilhelmsburg based on a valuation of their neighbourhood that is not only different but oppositional to its bad reputation. The very qualities that are considered "problematic," as I discussed in the previous chapter, are the basis of residents' valuation of it as a warm and welcoming place to live. Though *Heimat* is a concept with a troubled past and present, Wilhelmsburgers use it with comfort and warmth to indicate a place where they belong, feel comfortable, safe, and secure both as individuals and as racialized people in the context of a hostile world. With the term *Heimat*, and with their valuation of the neighbourhood's warmth, diversity, and acceptance of racialized people, Wilhelmsburgers express that the island is *home* in a context where their belonging is often questioned.

Figure 26: Houseboats on the Spreehafen, which until recently was inaccessible as it was part of the Freihafen (customs-free zone). In the evenings and on weekends many Wilhelmsburgers stroll, run, and lounge along its edge

(photo by author, 2017)

Heimat Wilhelmsburg

The portrait of the neighbourhood that Wilhelmsburgers offered me was rendered lovingly and in warm tones. The neighbourhood that they described has "its own charm" (Arzu, interview, 2017) – its green beauty, warmth, village-like quality, its diversity, the infrastructure that has been built up over years of migrant emplacement, and the reflection of racialized people in all aspects of neighbourhood life. This charm is the basis of the sense of belonging that emerged from so many of my interviews, where belonging is defined as a subjective feeling of rootedness and of being at home (Castañeda, 2018; Watt & Smets, 2014). The sense of knowing people in the neighbourhood and of being known appeared to be central to this sense of belonging. On the one hand this is likely an effect of long-time residence, which the evidence on place attachment demonstrates is the most consistent predictor of attachment to a place of residence (Lewicka, 2011). Certainly, Wilhelmsburgers described networks of friends, family, and acquaintances that had been established over time. Yet considering that the "village" of Wilhelmsburg in fact has a population of 55,074 people at last count (Statistisches Amt für Hamburg und Schleswig-Holstein, 2018), it is likely that public familiarity also plays a role. Public familiarity is a comfort with people that one sees and encounters but does not know personally, which develops through the use of a neighbourhood. It in turn contributes to one's sense of safety and ease in a particular space (Blokland & Nast, 2014).

Some of the qualities that Wilhelmsburgers love about the neighbourhood are associated with what Çağlar and Schiller (2018) call "migrant emplacement:" people's efforts to build a life within the barriers and opportunities afforded by networks of power and interconnection. The struggles for Muslim religious spaces in Wilhelmsburg are an example of this emplacement in process, as is the possibility of "getting by in Wilhelmsburg without German." Both are the products of people building a life for themselves, their families, and communities in the context of racial capitalism. As Wilhelmsburgers clearly emphasized, however, this emplacement cannot be comprehended only as *migrant* emplacement, though it is partly that. Wilhelmsburgers described their experiences of being *migrantized*, which is to say produced as immigrants when they are not (see El-Tayeb, 2016; Sow, 2018), despite being born and raised or settled in Germany over the long term. Often their own language and mine contributed to this migrantization, by using the racializing concept of people with migration backgrounds, but also "foreigners" and "migrants." Sometimes these terms are factually correct, or mobilized as political categories, in opposition to dominant uses of them to marginalize and exclude (cf. Çalışkan, 2011). Other times they represent the limits of language in a context where talking about race can be construed as racism in and of itself (cf. Barskanmaz, 2012).

In general, Wilhelmburgers' characterizations of the neighbourhood contradict the dominant reading of it from the outside. Residents' descriptions of the island as *diverse* in particular challenge the way in which the neighbourhood has been racialized and thus imagined as homogenous. This reflects the evidence that ethnically homogeneous neighbourhoods do not generally exist in Germany (Harlander, 2012); neighbourhoods that are constructed as problems are typically quite diverse (Keller, 2015). Rather what looks homogenous from the outside in fact includes various languages, religions, cultural backgrounds, and so on. The tendency to imagine neighbourhoods as homogenous is part of the racialization process itself, and as well as of the stigmatization of "Turks" and "Muslims" in Germany and the spaces with which they are associated (Çağlar, 1995, 2001; Ehrkamp, 2005, 2006; Güvercin, 2010; Ramm, 2010; Yildiz, 2009). As we will see in the following chapter, the actual diversity of Wilhelmsburg reveals the centrality of racialization to the "social mixing" strategies that the city has recently implemented in the neighbourhood.

In this context, Paul Gilroy's (2004) concept of "conviviality" helps to characterize what is important to Wilhelmsburgers while focusing on the processes of identification with each other and the island, rather than on the specific identities of each resident. Wilhelmsburgers describe history and practices of conviviality: the processes of living together and interacting that have made multiculture an ordinary feature of every life (Gilroy, 2004, p. xi). Its very ordinariness is a key part of what Wilhelmsburgers hoped to convey to outsiders. Crucially, conviviality does not mean an absence of inequity, it means that people are cohabitating and engaging with each other in the context of racism and intolerance (Gilroy, 2004).

When racialized Wilhelmsburgers talk about their neighbourhood and talk about their sense of belonging and being at home in it, they attribute positive value to the results of the concentration of racialized people in urban space. Spaces in which racialized people belong are typically stigmatized and pathologized, constructed as fundamentally outside of Europe, whether literally or figuratively (El-Tayeb, 2016). Within European cities this "outside" is construed as "the ghetto," "a patriarchal, violent, non-European space" that demands intervention and control (Stehle, 2006, p. 61). Migrant neighbourhoods in turn are construed as places that one should properly move *through;* to stay in a migrant neighbourhood for "too long" can be considered a failure to launch, and to properly use a neighbourhood as a springboard to other places (Harlander, 2012). Wilhelmsburgers, in contrast, demonstrate that there are benefits not to the racialization of space per se but to the safety, security, and comfort that a critical mass of racialized residents who have put down roots and invested in the neighbourhood can produce for those residents. Whether a neighbourhood is experienced as "safe" or "dangerous" in the context of stigmatization and pathologization of racialized space, and indeed the demonization and criminalization of racialized bodies, depends fundamentally on who is speaking. Visibility plays a crucial role here; residents described Wilhelmsburg as a place where racialized people, migrants, and Muslims are visible in general, and thus where *individuals* can experience a comfortable invisibility. Wilhelmsburg shares this quality with the notion of urban "enclaves," in which visual difference from "the mainstream" (i.e. white Germans) is minimized in the area, while the visibility of the enclave as a whole is increased (Hinze, 2013), as evidenced by the reputation discussed in the previous chapter. The island's diversity and its specific history prevents a definition of it as an enclave, however, which is typically thought of as a voluntary concentration of an ethnic group in a particular space (Hinze, 2013).

By calling the island *Heimat* in particular, Wilhelmsburgers are not mobilizing the entirety of its landscape of meaning, but rather are using it to convey an intersection of emotion, attachment, and place in the context of exclusion. The definition of *Heimat* that most fits how residents use it is the broad, open version proposed by Ursula Hildebrand (2003). In order to unsettle and draw attention to *Heimat's* baggage, Hildebrand frames it as an open question: where do you feel comfortable? That is *Heimat*, wherever or whatever it might be (Hildebrand, 2003). Yet the contested history of the concept and the very current present context throws into stark relief the significance of Wilhelmsburgers' assertations of belonging and being at home. As I established in the Introduction based on Fatima El-Tayeb's work, racialized people in Europe are produced as always out of place and out of time, as a result of their construction as outsiders to Europe and as supposedly behind Europeans in terms of development (El-Tayeb, 2016). This is reflected in the narratives used to racialize and exclude Wilhelmsburgers: the "guest" is temporary, whereas the German is imagined as permanent; the "foreigner" is out of place, while the German is at home; the person with "migration background" is marked by an arrival that is imagined as recent and definitive.

Yet racialized residents, some of them immigrants, feel themselves to be in place and at home in Wilhelmsburg. This echoes Çalışkan's (2011) findings in Berlin, that German-born people with Turkish roots assert a space of belonging within Germany through identification with the city and with the neighbourhood of Kreuzberg. Wilhelmsburgers use *Heimat* as a temporal and spatial concept (Clausen, 2008) to convey this belonging. In doing so they decentre the white, German subject of *Heimat*, and locate themselves as the subjects of this neighbourhood. I think of this as a process of "homing," of creating, establishing, and cultivating home (Boccagni, 2016), where "feeling at home is the opposite of feeling excluded or like an outsider" (Castañeda, 2018, p. 6), and is an accomplishment of rootedness and attachment in a context that is structurally hostile to the well-being of racialized people.

Even the concepts that are available to name something as home in German are exclusive and built on a racialized definition of belonging. Wilhelmsburgers' use of the term *Heimat* reflects a critical use of a concept that fundamentally excludes racialized people, immigrants, and Muslims, and that is often wielded for that specific purpose. Michel Foucault (2003, p. 267) defined critique as "the art of not being governed quite so much." When Wilhelmsburgers use the term *Heimat* to convey the intersection of belonging, identity, space, and emotion, they convey a sense of being at home that is critical of the attempted exclusions from belonging that I presented above in their words, and that is embedded in the language of *Heimat* itself. By claiming or perhaps acknowledging Wilhelmsburg as *Heimat*, they are not governed quite so much by the exclusion.

It is crucial to note that Wilhelmsburgers' use of the term *Heimat* does not reflect an idealization of their neighbourhood. The residents I spoke to are keenly aware of challenges, as I will discuss in detail in the following chapters. Some of those challenges are related to home in the sense of a physical place to live, as the availability of housing for low-income and racialized people is presently shrinking. Typically, residents talked about Wilhelmsburg's problems as part of their expression of it as *Heimat;* it was me who pulled the points apart for analytical ease. With Mostafa for example, I physically walked along the green canals and quiet fields that exemplify the central role that nature plays in *Heimat* in general (Blickle, 2002) and for Mostafa in particular. Yet as we walked, he told me about the commonplace conditions of poverty and poor housing on the island and walked me past the lot in the Korallusviertel that had gone fallow while the "property development sharks attacked a few times" (Mostafa, interview, 2017; see Figure 38). The beauty and attachment were not separated from what he saw as signs of neglect and political failure.

Further, as Costadura et al. (2019) argue, *Heimat* is not necessarily harmonious. To be *Heimat,* Wilhelmsburg does not need to be romanticized from the perspectives of the residents I interviewed, but rather more holistically seen and experienced. Complexity and disharmony, in keeping with Gilroy's definition of conviviality, is part of its charm. Conviviality does not exist in the absence of inequality and conflict, but precisely in that context, and the following chapters will further illustrate the inequalities and conflicts that are currently at play in the neighbourhood.

It is also the case that some of the green "nature" that residents value in Wilhelmsburg and that makes the island *Heimat* for them is the result of racial capitalism's shaping and exploitation of the landscape to maximize extraction. As I note in the captions to the figures for this chapter, the canals for example, which today afford pleasant if polluted views of tree-lined water, were dug to get industrial materials in and out of Wilhelmsburg as swiftly as possible. As such they might be thought of as material traces of the structural past and present, particularly as they are now being re-framed as spaces for leisure and art in the interest of the middle class, as I will discuss in the following chapters. Further, as my photographs demonstrate, the industrial side of Wilhelmsburg is always around the corner and sometimes pressed right up against the green and the beautiful, and thus one is always in view of environmental racism, of pollution, and of the racialized economic system. Certain green spaces like the Spreewald were specifically produced by the processes of abandonment that the previous chapter detailed. The small forest, which is located across from where I lived in Reiherstieg, has grown since it was abandoned by the city after the 1962 flood. The young forest did not come up in my interviews, but is the focus of resident activism to "save" it from redevelopment into housing in the next phases of city intervention on the island (Waldretter, 2019).

In the photos of old trees with which I have bookended this chapter I play with the disharmony that is present on the island (Figures 13 & 27). The first, taken on

Rotenhäuser Damm, looks beautiful and lush. Hidden from view is its context in an apparently abandoned lot that is fronted by a small sign that advises passersby to keep out. During the International Building Exhibition, which I will discuss in the next chapter as a city strategy to redevelop Wilhelmsburg, the site was a stage for a partnership between the development project and a local university through which students, scholars, artists and urbanists were attracted to the island (IBA Hamburg, 2020). The lush green is at once beautiful and inaccessible, and symbolic of state efforts to transform the island away from how the Wilhelmsburgers I spoke to know and value it now. The tree at Haulander Weg is likewise in one of the beautiful green spaces that I appreciated and admired in the time that I spent on the island. The old fruit tree stands in a small untended grove that I photographed when it was carpeted by spring flowers. I took the photo, however, on a bike tour of planned sites of redevelopment on the island and framed the photo in such a way that the highway running past just metres away was at my back and out of sight. With both photos I offer a deliberately selective view of disharmonious and conflictual places and moments in time. These places contrast with the ordered green space that has been produced in recent years, as the next chapter will address.

Finally, I found that Wilhelmsburgers have the kind of longing that is typically associated with *Heimat* in two senses: they long to *remain* at home in their *Heimat* in the context of redevelopment and change, and they long to live well and more closely with the full diversity of their neighbours. In the following chapters I will first present the urban planning interventions that are now shaping *Heimat* Wilhelmsburg according to the logic of "social mixing," and will then turn again to Wilhelmsburgers' perspectives, hopes, and fears in relation to these changes.

Conclusion

In this chapter I traced out the image of Wilhelmsburg that the racialized long-time neighbourhood residents I interviewed felt was often neglected or missed in media, research, and other outside views of the island. After introducing the concept of *Heimat* that residents often use to describe the island, I presented what they love about it, from its green and beautiful landscape, to the warmth and friendliness that they experience with neighbours, and its village-like quality, which they distinguish from other urban spaces in which people are anonymous. In Wilhelmsburg, people know each other, you are recognized and appreciated, while at the same time you can blend in. Racialized people are present in all aspects of daily life in Wilhelmsburg, which makes it different from other neighbourhoods where racialized people feel exposed, scrutinized, and subject to the question "what are *you* doing here?" I characterized this quality as conviviality (Gilroy, 2004), the process of cohabitation and engagement that makes multiculture ordinary.

Figure 27: Aged fruit tree at Haulander Weg, a spot slated for redevelopment in the next few years

(photo by author, 2018)

In contrast to the homogenizing, racializing bad reputation of the island, Wilhelmsburg is diverse, and people with various backgrounds and socio-economic statuses live together with a degree of familiarity that long-time residents value. I argued that much of what Wilhelmsburgers appreciate about the neighbourhood has been cultivated through years of migrant emplacement in the context of the neighbourhood's devaluation and stigmatization. But one person's "ghetto" is another's *Heimat,* and Wilhelmsburgers communicate a striking depth of emotional attachment to their neighbourhood. In this chapter I have explored the relationship of these perspectives on the neighbourhood to the kinds of discrimination and exclusion that Wilhelmsburgers experience as immigrants and racialized people in Germany in general. I argued that the neighbourhood is special to residents because of the exclusions from Germanness that they describe experiencing. In Wilhelmsburg, they belong, whereas they struggle for that belonging in the nation. I closed the chapter with my reflection on Wilhelmsburgers' perspectives, arguing that they demonstrate a notable sense of belonging and of place attachment, which as Nalan argues can be viewed as a strength. Remarkably, Wilhelmsburgers take up the concept of *Heimat* to assert their sense of being at home in the neighbourhood, a concept that I demonstrated is a racialized, exclusive concept of belonging, identity, and space. By doing so, they enact their own critique of the exclusion that they otherwise experience and assert a space of racialized and migrant belonging.

In the following chapter I turn to the research findings that emerged from my reading of Hamburg policy and media archives, as well as from interviews with residents, politicians, and planners. I introduce the urban planning interventions that the city-state of Hamburg has implemented in Wilhelmsburg since the early 2000s, and situate them within racialized residents' sense of attachment and belonging and within the longer history of development on the island. With this structure I continue to follow the direction of the residents I interviewed, for whom the city-state's highly publicized event-projects were not necessarily as impactful or deserving of attention as they appeared (to me) from the outside. I emphasize the policies and developments that residents found to be of some significance and add to that the insights of local researcher Zeynep Adanalı (2013), whose master's research identified housing policy that is not well publicized. The city-state's interventions are based on a logic of "social mixing," in which the island is made attractive to middle-class, white Germans. I explore critiques of this as state-led gentrification, arguing that this lens does not account for the specifically racial strategies in which the city is engaged. I argue that through social mix strategies the city produces racialized "displaceability" (Yiftachel, n.d.) that is consistent with the long-standing devaluation of the island and its residents, and that risks un-homing the very residents who have created Wilhelmsburg as their *Heimat* over decades and generations.

5. Planning Strategies to "Restructure" Hamburg-Wilhelmsburg
Mix and Mega-Projects

Introduction

As a city bus rolls through the centre of Wilhelmsburg, it passes by a cluster of colourful new buildings. On the bus, a passenger affects the tone of a tour guide for the benefit of a camera: "Ladies and gentlemen!" they announce, "We're now driving by Legoland! Here we have the new Disneyland... We want new people in Wilhelmsburg, and this is just the thing!" (in Schaefer, 2013a, translation by author). This is a scene from *Die Wilde 13* (the "Wild 13"), a video ethnography of the eponymous bus route, which I picked up along with an accompanying book (Schaefer, 2013b) during my archival research at the *Geschichtswerkstatt Wilhelmsburg & Hafen*. The ethnography approaches the bus that traverses Wilhelmsburg from north to south and all points in between as a microcosm of the island, a place where one encounters the diversity of Wilhelmsburg in an enclosed space. Along the way, "the past and future pass by its windows," offering a view onto the island where "the gentrification carousel has begun to turn" (Schaefer, 2013a, n. on DVD version, translation by author).

While Wilhelmsburgers' opinions vary about whether and how the neighbourhood has changed in recent years, as I will discuss in the next chapter, what is clear is that there has been a shift from Wilhelmsburg as a stigmatized, beloved neighbourhood with a bad reputation to a place that the city is trying to "hype up," (Mohammed, interview, 2017) and that has even "become a bit trendy" (Murat, interview, 2017). As Umut put it "for years, people forgot that Wilhelmsburg was two stops away from the central station [*laughs*]. For years. And then at some point they looked at the HVV [public transit] map: 'Hey! That's not so far after all! [*laughs*] Now let's do something here'" (Umut, interview, 2017).

In this chapter, I describe the "something" that has been done to date by the city-state of Hamburg. Since 2005, the city has implemented a subsidy for student housing, a social mix policy in public housing, and the twin event-projects of the

International Garden Show and International Building Exhibition. All of these interventions were informed by the myth (Holm, 2009) and mantra (Münch, 2014) of social mix as a solution for Wilhelmsburg's problems, and aimed to set off a shift in outside perceptions of the neighbourhood and thus to attract the (white) middle class. The following chapter will explore residents' perspectives on what, if anything, these strategies have achieved in Wilhelmsburg thus far. This chapter focuses on the official logic of each of the interventions that were identified by study participants – residents, politicians, and planners – as central to their recent experiences and to the city's plans for the island. The social mix policy is an exception, as it did not come up in my interviews, but was identified by local researcher Zeynep Adanalı (2013). In this chapter I outline each of the interventions based primarily on the archival documents of agencies of Hamburg city-state.

This chapter also situates recent planning interventions as contested. I draw on the work of local scholars and activists, media archives, interviews with racialized residents, politicians, and planners, and on participant sensing in the neighbourhood to highlight protest and debate in Wilhelmsburg. Though Hamburg city planning adopted as its motto "*Aufwertung ohne Verdrängung*" (improvement without displacement), local scholarly and activist work has emphasized the threat of displacement of low-income residents and critiqued the production of space for visitors and the middle class. Residents have also critiqued the "resident engagement" processes that have been part in particular of IBA and IGS developments as strategies for producing acceptance rather than participation. While some analyses of development in Wilhelmsburg have touched on planning as governmentality (Dörfler, 2014) and on mixing as a form of "population politics" after Foucault, which governs at the level of population and bodies (Hohenstatt & Rinn, 2013; Rinn, 2018), racialization has not been part of the analyses captured in the media or in academic scholarship. The gap that this research has aimed to fill is thus especially visible here.

As an analytical frame, gentrification does not fully capture the racialized nature of displaceability – the susceptibility of particular people and groups to being "removed, expelled or prevented from exercising their right to the city" (Yiftachel, n.d.) – that is produced by planning strategies in Wilhelmsburg, and particularly by the logic of social mix. The common sense of social mix in German urban planning masquerades here as solely class-focused, while the realities of how it is implemented and monitored illustrate the central concern with control and distribution of people who are racialized and impoverished within racial capitalism. The most prominent criticism of city-state interventions in Wilhelmsburg is that they are likely to produce displacement over time. Displacement must be understood as a function of racial capitalism and a form of state violence in which racialization plays a critical role, one that the lens of gentrification does not pinpoint completely or with precision (Roy, 2018). The production of displaceability in Wilhelmsburg demonstrates the devaluation of racialized residents and of racialized *Heimat* as the previous chapter explored. Through the racialized and classed production of displaceability, the policies of the Hamburg city-state threaten the processes of emplacement, attachment, and spatialized identification, putting racialized residents at particular risk of "un-homing" (Elliott-Cooper et al., 2020).

I begin by briefly describing a turning point in Wilhelmsburg, the period in which, according to many of the residents and local politicians with whom I spoke, the city-state was forced to change tactics on the island from neglecting it and using it as a dumping ground as the dominant mode, after a period of political instability in which many Wilhelmsburg voters moved sharply to the right. I discuss the myth of mix that underlies the city's change in strategy. The logic of counting and sorting bodies is revealed in particular through social monitoring, which postdates the earliest interventions I describe here, but is useful for understanding the underlying logic. I then describe three key interventions that ensued, starting with the policy changes and then introducing the event-projects. I highlight some of the debates captured in the media about the event-projects in particular, and discuss critiques of these interventions as gentrification, particular from scholarly and leftist activist circles in Wilhelmsburg and Hamburg. IBA Hamburg's Global Neighbourhood was a major focus of debate, although – or perhaps because – it was exceptional among the IBA's many projects. I conclude the chapter by highlighting the presence of racialization in the city-state's policies and projects, guided by Ananya Roy's (2018) question "who can count as a subject, who can claim home and land?" As I signalled in Chapters 1 and 3, the answers to this question are embedded in the city's ongoing devaluation of racialized residents, in keeping with its colonial legacy and investments in racial capitalism.

The instruments of restructuring

Urban panics about neighbourhoods considered "problems" like Wilhelmsburg tend to mobilize state resources, even as they solidify the stigmatization of racialized people (Tsianos, 2013). Hamburg-Wilhelmsburg entered the twenty-first century in a storm of negative media attention. The long-standing bad reputation that I detailed in Chapter 3 reached a fever pitch around the turn of the century. In a nationally televised debate between candidates for the German Chancellorship, Wilhelmsburg was mentioned as an example of a "failed neighbourhood" (Weinreich, interview, 2018). It was labelled as a ghetto again and again in media coverage of several tragic deaths that took place in the summer of 2000. A "cry for help from the Bronx of the North" was issued by some Wilhelmsburg social workers, doctors, teachers, and politicians, in the form of a community event, in which they argued for an urgent injection of resources by the city-state (Hamburger

Abendblatt, 2000; Meyer-Wellmann, 2000). The sense was that Wilhelmsburg had had "zero support" for decades, as Arzu put it, and according to Michael Weinreich (SPD) – a Wilhelmsburger born and raised and presently the island's representative in Hamburg parliament – "that was the mood in Wilhelmsburg: 'We're fed up'" (Weinreich, interview, 2018).

With what in particular residents were fed up varied. Some Wilhelmsburgers were fed up with a lack of government resources on the island and with continued attempts by the city to locate undesirable functions there, such as highways and incineration facilities (see Humburg & Rothschuh, 2018; Zukunft Elbinsel Wilhelmsburg, 2012). Concerns about perceived out-movement of middle-class, German families also frequently appeared in the activism of the day, as Serdar (interview, 2017) touched on when he mentioned that in the 1990s "many, many young Germans left or sent their kids to other schools, to other parts of town." Some people were concerned about the condition of schools on the island, and young families with kids in particular decided to leave rather than send their children to Wilhelmsburg schools.

Perceptions of this out-movement, which I understand as "white flight" (cf. Andersen, 2017; Araujo, 2016; Bråmå, 2006), and interpretations of its meaning varied among my interviewees and others I met in Wilhelmsburg, as I will discuss in the following chapter. At the time the right-wing "law and order" Schill Party, a protest party which was formed in 2000 by a Hamburg judge (Faas & Wüst, 2002), found traction on the island "with the sole motto: 'we have too many foreigners'" (Weinreich, interview, 2018). Wilhelmsburg was historically a Social Democratic (SPD) stronghold, and a bastion of left-wing and Communist politics prior to the National Socialist era (Behörde für Schule und Berufsbildung, n.d.; Geschichtswerkstatt Wilhelmsburg und Hafen, 2018). Yet in the 2001 Hamburg elections, the Schill Party won a whopping 34.9% of the vote in Wilhelmsburg and 19.4% of the vote Hamburg-wide (Faas & Wüst, 2002; Statistisches Landesamt der Freien und Hansestadt Hamburg, 2001). Crucially for Wilhelmsburg (and beyond), only people with German citizenship can vote, including at the level of the city-state. Even since 1999 reforms, restrictions on eligibility for citizenship limit access to voting rights, especially for people who want to maintain citizenship in another country, who are unemployed, who have criminal records, or who are not fluent in German (Bucerius, 2012; Çalışkan, 2011; Institut für Migrations- und Rassismusforschung Hamburg, 2011). Thus, many Wilhelmsburgers are excluded from the process, creating what some of my interviewees indicated is a fundamental problem with political representation on the island.

The 2001 election was a turning point in Wilhelmsburg's relationship to the city according to several of the local politicians I interviewed. Though the Schill Party would collapse in two short years, the vote was a wakeup call for mainstream political parties in the city in general, as the SPD lost power for the first time in

44 years (Faas & Wüst, 2002). The city-state could no longer ignore Wilhelmsburg or treat it simply as the place where "anything we don't want in Hamburg goes" (Heike Sudmann, *Die Linke*, interview, 2018). One result was the constitution of the Wilhelmsburg "Conference for the Future" (*Zukunftskonferenz*), a group of 100 resident volunteers and representatives of local government who came together to produce a roadmap for the future of the island. Their 200-page report produced by seven working groups argued that it would be a shame to continue along a path of "slum and ghetto development," "stacks of container mountains," and planning as though the port would be king forever (Zukunftskonferenz Wilhelmsburg, 2002, p. 6, translation by author). Instead the city should "create the conditions for future-oriented development at the heart of Hamburg" (Zukunftskonferenz Wilhelmsburg, 2002, p. 6, translation by author).

This "white paper" would to some extent inform the city's plans for development in Wilhelmsburg. Michael Weinreich (SPD), who represents the island in Hamburg parliament, explained the core messages in terms of implementing plans and policy to attract new people to the island, while protecting the long-time population from displacement.

> A demand was always – we had the Future Conference, the White Paper, each step led to another and then to the IBA. It was always in-migration and new builds in Wilhelmsburg, to have a mixing. There are two options: either I gentrify it, so that half of the people have to leave the neighbourhood, or I build onto it and thus have the possibility of changing it without expelling everyone [*laughs*]. So, obviously to have in-migration, the socially acceptable option is – and I think it is the right way, too – it's what's happening now. We're going to build over 5,000 apartments, which mean that 10,000 people will move in. That will be a big change here (Weinreich, interview, 2018).

To achieve this in-migration, the city planned three prongs of policy and development projects; the following section describes these three prongs.

The guiding myth of social mix

At the centre of the three prongs of city-state intervention is the "myth" of social mix, which, as I established in Chapter 1, is a long-standing common sense in German urban planning and beyond (Bayer et al., 2014; Holm, 2009; Münch, 2009). It is referred to as a myth, an ideology (Kuhn, in Wehrheim, 2018) and an intuition-based mantra (Münch, 2014) by its critics, because it is based on a shaky evidentiary foundation and yet is repeated over and over as a guiding principle and practice, regardless of outcomes (Bridge et al., 2012). The idea of social mix has featured in German urban planning since at least the 19th century, when it was considered a tool for dissolving unmanageable neighbourhoods and controlling through dispersal people who were likely to be frustrated by unemployment and poverty (Holm, 2009). Today it remains a common sense, though there is often a lack of clarity regarding what problem(s) social mixing is meant to solve (Harlander & Kuhn, 2012), and what kind of mix is desired and why (usually income, ethnicity, and/or tenure) (Saville-Smith et al., 2015). I introduced the notion of social mix as a strategy of dispersal of racialized people in Chapter 1, and here I will briefly discuss how social mix is anchored in Hamburg in the guiding policy for new developments and in the practice of "social monitoring."

Firstly, it is useful to have a sense of the overall structure of housing in Hamburg. As is generally the case in Germany (Münch, 2010), it is much more common to rent than to own one's housing in the city. As of 2020, just over 76% of housing in Hamburg was rental housing, and 23.9% was owner-occupied (Krieger et al., 2021). Likewise 80% of housing was in multi-unit buildings and just 20% in buildings with less than three units, according to the most recent available data (Adanalı, 2013). The vast majority of housing in Hamburg is provided by small, private landlords who rent out just a few apartments, and this matches the countrywide pattern (Adanalı, 2013; Münch, 2010). In general this means that small, nonprofessional landlords hold a great deal of power in the housing market. In Wilhelmsburg, which has a high percentage of social housing in comparison to other parts of the city, the various institutions that provide that housing, including municipal corporations and cooperatives, also play a strong role (Adanalı, 2013).

Social housing comprises a significant part of the rental housing structure in Hamburg and in Germany more broadly. In 2016, 8.4% of housing in Hamburg was social housing (Statistisches Amt für Hamburg und Schleswig-Holstein, 2018), which is to say that the rent was controlled and publicly subsidized, with rental limited to households with incomes below a certain level (Münch, 2010). The percentage of social housing is currently on a sharp downwards trajectory, which is part of a nationwide pattern. In recent years, social housing has been cut in half: in 2003, 16.4% of housing in Hamburg was social housing (Statistisches Amt für Hamburg und Schleswig-Holstein, 2006). Across the country, the dedication of units to subsidization expires in 15- to 30-year intervals, and at the same time fewer (if any) new subsidized units are created (Zeit Online/dpa, 2019). Münch (2010) indicates that this "expiry" is a uniquely German formulation, which has been created over decades of liberalization of housing policy. Since World War Two, when 20% of West Germany's housing stock was destroyed, national policy has aimed at supporting the building of housing yet deregulating and encouraging ownership at the same time. In the 1990s the formal status of housing shifted from being considered a social good, to an economic product (Münch, 2010).

In Hamburg, "people with migration backgrounds" are more likely to rent their housing (86.1%) than are "people without migration backgrounds" (76.6%) (Statistisches Amt für Hamburg und Schleswig-Holstein, 2013). Further, Adanalı (2013) reports that a majority of Muslim Hamburgers live in social housing, as do slightly less than half of residents who have roots in Turkey. This is higher than the 14% of Germans who had subsidized housing in Hamburg around the same time, and higher than the 14.3% of people with roots in Turkey who lived in social housing across Germany (Adanalı, 2013; Friedrich, 2008). Differences in the social housing rates of various migrant communities are generally attributed to the combination of when they arrived in the country and what the specific policies were at the time (Friedrich, 2008; Schönwälder & Söhn, 2009). Until the late 1970s, people without German passports were excluded from renting publicly subsidized apartments (Schönwälder & Söhn, 2009).

Wilhelmsburg had an unusually high proportion of social housing at the introduction of the policies and projects discussed in this chapter. In 2003, 37.6% of housing was subsidized on the island (Statistisches Amt für Hamburg und Schleswig-Holstein, 2006). While there is a lack of data about the specific demographics of residents of social housing in Wilhelmsburg, the map of social housing concentration matches the map of where residents with "migration backgrounds" are also most concentrated (Adanalı, 2013, pp. 32 & 43). This is particularly in the areas of Reiherstieg, Bahnhofsviertel/Korallusviertel/Schwentnerring, and in the cluster of highrises in Kirchdorf-Süd. By 2016, subsidized housing had decreased to 23.9% on the island, with the subsidization of a further 13.1% of that set to expire in 2022 (Statistisches Amt für Hamburg und Schleswig-Holstein, 2018).[1]

In recent years, Hamburg has introduced a new policy instrument to encourage the "mixing" of different kinds of housing in the city. According to the "Hamburg Mix-of-Thirds," a new standard for development in Hamburg (Vogelpohl & Buchholz, 2017), mix in tenure and income is required for any new housing units built in the city. A third should be owner-occupied, a third market rental, and a third publicly subsidized rental units (Freie und Hansestadt Hamburg, Senatskanzlei, n.d.). The politicians I interviewed suggested that this was a rather elegant and practical solution to the perennial political challenge in Wilhelmsburg and elsewhere: that some people argue "we need social housing" and others "we need single family homes." Heike Sudmann, who is the speaker in Hamburg parliament on housing, transport and development issues for *Die Linke*, assessed the notion of mix in the following way: "I would say that it was a good *idea*. Yes, it was always actually conceived from the perspective of those who have fewer opportunities on the housing market or wherever, that they could also get into the neighbourhoods where otherwise only the well-off live. So, it was conceived as '*you* have to open up too'" (Sudmann, *Die Linke*, interview, 2018). Hakverdi (SPD, interview, 2018) likewise called it a "politically brilliant construction." He told me that the debate over different

1 As of 2020 the percentage had fallen slightly to 23.6%, now with 19% that is set to expire in 2026 (Statistisches Amt für Hamburg und Schleswig-Holstein, 2020).

kinds of tenure was long-standing in Wilhelmsburg, and that the tone of the debate depended on what part of the island one went to. I heard a lot of demand for affordable housing, as I discuss in the following chapter, but Hakverdi suggested that if I went to the residents' association in Kirchdorf I would hear something quite different: "There they will say: 'Not a single new social housing unit, please. *Niente.* Single family homes. We want German families to move here'" (Hakverdi, interview, 2018). This illustrates the political debate about what to build where, but also the muddling of the terms of mixing, where social housing is considered to be representative not only of income and type of tenure, but also of race/nationality.

Underpinning the notion of social mix in German urban planning today is the concept of neighbourhood effects, wherein it is argued that low-income neighbourhoods themselves present a problem for poor and marginalized residents. The evidence from German cities does not support this claim (Wehrheim, 2018), and elsewhere is also unclear at best, but policymakers and researchers demonstrate significant *belief* in neighbourhood effects and their importance (van Ham, 2012). One place where this is evident in Hamburg is in the politics of "social monitoring," which assumes that the presence of "social problems" in a particular space itself has negative effects on residents (Pohlan, 2018). Every year since its inception in 2009, following a model implemented a decade earlier in Berlin (Pohlan, 2018), the city's social monitoring is undertaken as an "early warning system to identify neighbourhoods with cumulative critical issues," where the city-state can then take local, spatial action (Behörde für Stadtentwicklung und Umwelt et al., 2010, p. 9, translation by author; see also Freie und Hansestadt Hamburg, n.d.-a).

Social monitoring uses markers of class and the racializing concept of migration background, which I have discussed extensively in previous chapters, as indicators of potential problems. The analysis uses seven demographics to assess and assign a status to every part of Hamburg, at the level of census tracts (Behörde für Stadtentwicklung und Umwelt et al., 2010). The status of an area is defined based on the percentage of certain groups that reside in it and whether those populations have grown, shrunk, or stayed the same since the previous year. The demographics that are used to make this calculation are: children of single parents, people who have left school without a diploma, unemployed people, children and elders living on social assistance, and children and youth with so-called migration backgrounds (Behörde für Stadtentwicklung und Umwelt et al., 2010).

The originators of the monitoring tool, a partnership between academics from local universities and the Hamburg Ministry for Urban Development and Environment (now Urban Development and Housing), argue that in the social monitoring, the demographics themselves stand in for social inequality and disadvantage; the monitoring tool counts the marginalized in order to identify the concentration of marginalization. However, the problematization of *concentration* rather than of marginalization itself, alongside the counting of racialized and classed bodies as the main form of measurement, adheres the "problem neighbourhood" to residents themselves. It also reproduces the stigma attached to unemployment, receiving social assistance and so on. This is thus one of the ways in which the blaming of individuals for structural problems, which was an important theme in the previous chapters, is concretely operationalized.

Social monitoring began in Hamburg when the interventions in Wilhelmsburg that I will describe below were already underway, and I therefore detail its functioning not to imply a causal connection, but rather to illustrate the dominant logic of mix. The logical counterpoint to the definition of concentration of low income and racialized people as problems is that, in turn, "the moving in of higher-income earners and of educated middle-class German families supposedly makes neighbourhoods better" (Holm, in Bayer et al., 2014, p. 8, translation by author). As Michael Sachs, former housing coordinator of the Hamburg Ministry of Urban Development and Environment, put it in a volume on social mix practice in Germany:

> Essentially, it's about choosing, across ethnic and economic boundaries, socially-competent renters who are able to pay the rent, and who have the recognizable skills for neighbourly life. If one understands these competencies as middle-class virtues, then it's about bringing middle-class elements into difficult neighbourhoods (Sachs, 2012, p. 409, translation by author).

Wilhelmsburg academics and activists Florian Hohenstatt and Moritz Rinn (2013) astutely identify this as population politics after Foucault, wherein governance at the level of the population (biopolitics) is achieved by control and exclusion to change "mix." In the mix is a combination of moral governance by apparently bourgeois virtues. It is not only about money but about "skills" and thus the instruction of those perceived as unskilled. The claim that Sachs advances, that the selection process cuts across class and race, depends upon disregard for "competencies" as a culturized and racialized notion (see for example earlier discussion of integration). In any case the evidence contradicts the claim: the mixing of Wilhelmsburg is explicitly along boundaries of economic and "migration" status.

As a state strategy that attempts control of racialized and poor people through a combination of dispersal and proximity (Kipfer & Goonewardena, 2014), social mixing functions in Wilhelmsburg as a specifically "racialized technique of recasting relations of domination" (Kipfer & Goonewardena, 2014, p. 103). While Sachs acknowledges that the idea that one can plan a somehow "better" mixture of people has a certain "off taste" given the historical intersections of notions of mixing and hygiene in Germany (Sachs, 2012, p. 409, translation by author), discussion of social mixing based on "ethnicity" actually does not carry the same taboo in Germany that it does in France, for example (Kipfer, 2016). On the contrary, ethnic mix is considered a valid measure against segregation, and the right of housing providers to restrictive allocation procedures is built into national housing law on that basis (Bolt, 2009; N. Ha, 2014b; Harlander, 2012).

Though Germany has a General Law on Equal Treatment[2] that aims to "prevent or eliminate discrimination based on racial or ethnic origin, gender, religion or belief, disability, age or sexual identity," there is an exception to the law when it comes to housing (Bundesministerium der Justiz und für Verbraucherschutz, 2006). Paragraph 19 (3) establishes that: "in the case of rental housing, a difference of treatment shall not be deemed to be discrimination where they serve to create and maintain stable social structures regarding inhabitants and balanced settlement structures, as well as balanced economic, social and cultural conditions" (Bundesministerium der Justiz und für Verbraucherschutz, 2006). The legality of this exception was challenged by the EU, but was eventually ruled to be consistent with European legal standards (European Commission, 2010; Münch, 2009).

In the exception to the law on equal treatment, there are no definitions of the notions of balance, stability, structures, or conditions, which allows for their subjective interpretation and thus a quasi-official implementation of quotas for certain kinds of people in public housing (Adanalı, 2013; N. Ha, 2014b; Harlander & Kuhn, 2012). The premise is the disruption of homogeneity in urban spaces, though it does not generally exist along ethnic lines in Germany, but rather in terms of spatialized poverty and racialization. In any case, it is white German communities that are much more likely to be homogeneous, while neighbourhoods like Wilhelmsburg are in fact quite diverse; the segregation that does exist is caused by the dominant population, as I discussed in Chapter 3 (Harlander, 2012; Wehrheim, 2018, Kuhn, in 2018). Yet it is Wilhelmsburg that has a high – which is taken to mean problematic – social monitoring score (Pohlan, 2018).

What I will argue is that social mix policies and practices as they have recently been implemented in Wilhelmsburg produce specifically racialized displaceability that aligns with racial capitalism's sorting and devaluation of certain bodies (Pulido, 2017). Hamburg city-state is using displacement as a strategy to produce space that aligns with a bourgeois ideal (Hohenstatt & Rinn, 2013; Rinn, 2018), and this must be understood as a specifically white, German bourgeois ideal. Racialized residents who have made a home in Wilhelmsburg and who are deeply attached to the island are at risk of displacement as an embodied and racialized form of state violence (Roy, 2018), particularly if they have low incomes, precarious immigration status, or are unemployed or criminalized. Thus as Yiftachel (n.d.) argues, displacement must be viewed within histories of colonialism, slavery, and imperialism, and the "identity regimes" that persist from them, spheres of power "where the status, resources and visibility of groups are determined, negotiated and challenged."

2 The *Allgemeines Gleichbehandlungsgesetz* came into effect in 2006 in response to European anti-discrimination directives (see Federal Anti-Discrimination Agency, 2010).

In the following section I outline the core interventions that Hamburg city-state has initiated and unrolled in Wilhelmsburg since 2004, before turning to some of the local debates that have arisen in response to these interventions.

Subsidies for student housing

The earliest intervention that long-time residents described to me was a subsidy program introduced by the city of Hamburg to encourage students to move to Wilhelmsburg. Beginning on the neighbouring island of Veddel in 2004 and expanding to Wilhelmsburg in 2006, the program offers an opportunity for students and trainees to pay below-market rents, if they are signing new rental agreements in particular geographical areas. Through the program, students pay a limited rent (€210 in Wilhelmsburg) and the state offers landlords a top-up and security guarantee. At first the subsidies applied particularly to the Reiherstieg area of Wilhelmsburg, and later apartments at the centre of the island were eligible, in the area around the Wilhelmsburg S-Bahn station and Luna Center mall (Berta-Kröger-Platz) (Hamburgische Investitions- und Förderbank, 2016). The program applies only to apartments with more than two bedrooms and thus encourages the creation of shared student housing (Behörde für Stadtentwicklung und Wohnen, n.d.; "Subventioniertes Wohnen," 2013).

According to the city-state of Hamburg, the logic of the subsidies is twofold: they provide affordable housing for students and they "support neighbourhood development goals" (Hamburgische Investitions- und Förderbank, 2016, p. 3, translation by author). As the institution responsible for disbursing state funds puts it, through the subsidy program "a contribution can be made to changing the resident structure, and alongside other neighbourhood development measures, can support development into an interesting and sought-after residential quarter" (Hamburgische Investitions- und Förderbank, 2016, p. 3, translation by author). When the subsidy was introduced in Wilhelmsburg, Hamburg's Senator for Urban Development and the Environment at the time, Michael Freytag, claimed that "we support student living because it strengthens the neighbourhood individually. Those who choose the Elbe island of Wilhelmsburg will soon be living in the front row, as this is where the Leap Across the Elbe begins" (quoted in Rebaschus, 2005, translation by author).

The Leap Across the Elbe (*Sprung über die Elbe*) is the city of Hamburg's framework for development towards the south of the city, with the island of Wilhelmsburg at its core. Nested within a long-term plan to pursue population growth for the city (Freie und Hansestadt Hamburg, Staatliche Pressestelle, 2002), the Leap Across the Elbe was conceptualized as a literal about-face in Hamburg planning (Behörde für Stadtentwicklung und Umwelt, 2005). Whereas development of the city had long been oriented northwards from the river, the Leap Across the Elbe turned to face southwards towards the Elbe islands, with a vision of concentrating development closer to the city's core (Behörde für Stadtentwicklung und Umwelt, 2005).

This turn southwards had in effect already begun with the development of HafenCity, which I touched on in Chapter 3. Beginning in 1997 and continuing to expand today, HafenCity is a completely new neighbourhood built immediately south of the historical city centre on port lands that had become disused. It has been criticized as a purpose-made, homogeneous, rich enclave that attracts wealthy residents and high-status businesses through its novelty and location, which only the wealthy can afford (Menzel, 2010). Its overall homogeneity shifted slightly with the addition of a container village for refugees in 2016. This developmentwas distant from established parts of HafenCity, however, and in any case was temporary, slated for closure in 2020 to make way for imminent, high-end redevelopment (cf. Baumgartner, 2019; Marinov, 2016).

With the plan to continue southwards along a "development axis" between HafenCity and Harburg, the Elbe islands – and Wilhelmsburg in particular – became the "most important area of development for the city" (Walter, 2005, p. 72, translation by author), and multiple policies, projects, and investments were planned accordingly.

"Mixing" of publicly funded rental housing

Beginning in 2005, Hamburg implemented a policy of *Durchmischung* (intermixing)[3] in publicly owned rental housing in Wilhelmsburg. In contrast to the student subsidies and mega-projects, this practice is not reflected in official policy frameworks or documents, but rather was identified through the work of local researcher Zeynep Adanalı (2013). Adanalı studied the experiences of people with Turkish backgrounds in the housing market in Wilhelmsburg. She found that independent of their economic status, people with Turkish backgrounds who were looking for rental apartments faced intense difficulties specifically accessing publicly owned apartments. Adanalı traced the difficulties back to instructions to public housing authorities from the Christian Democratic Union (CDU) government in 2005.

In that year, public housing corporations, which together account for over 50 per cent of housing in Wilhelmsburg, were instructed to "reach out to socially-

3 While *Durchmischung* is most often translated as 'mix,' the English word does not quite convey the full sense of the German, which has more of an assimilationist connotation. When used in the context of chemistry and biology for example, *Durchmischung* is also translated as diffusion, dilution, or stirring, and some of this connotation is present in the term's use in planning, social sciences, and public discourse.

stable families who do not have a migrant background," in order to improve the social mix of the neighbourhood (Adanalı, 2013, p. 97, translation by author). In 2007 the city also made Wilhelmsburg an area of exemption to upper limits on income for renters of publicly funded housing (this continues until 2025, see Bürgerschaft der Freien und Hansestadt Hamburg, 22. Wahlperiode, 2021). Adanalı emphasized, however, that in practice even the applications of higher-income people with "migrant-sounding names" were being set aside in the name of "mixing" (Adanalı, 2013).

The policy that Adanalı uncovered was not driven by formalized criteria, but by housing workers' subjective assessments of the threat of "ghettoization" in Wilhelmsburg, or indeed in particular buildings and blocks of housing (Adanalı, 2013). This is in fact consistent with the law, as I introduced above. The policy is little-known in Wilhelmsburg though it is strongly felt by racialized residents, as I will demonstrate in the following chapter.

The subsidy for student housing and the social mixing directive set the stage for the two major projects that began in 2006 and culminated in 2013 events: the Hamburg International Building Exhibition (IBA Hamburg)[4] and the International Garden Show (IGS). These projects were at the centre of development strategy for the island, intended to mobilize resources and concerted effort from the government and its ministries, and thus to serve as engines for the Leap Across the Elbe (Walter, 2005). The projects comprised a "positive state of exception," as it was described to me by a local planner (Interview, June 6, 2018). In this state of exception, inter-ministry cooperation was heightened, and development sped along by the loosening of regulations that would ordinarily slow development processes. This is typical of IBAs in general (Humburg, 2009). The Ministry of Urban Development and Environment had also hoped for two other events to drive the Leap – the naming of Hamburg as Cultural Capital of Europe (2010) and the Olympics (2012) – but neither bid was successful. In 2015 Hamburg residents further voted against bidding for the 2024 Olympics as well, taking that particular development strategy off the table for the time being (Vogelpohl & Buchholz, 2017).

International Garden Show

IBA Hamburg and IGS were separately initiated but coordinated undertakings as part of the Leap Across the Elbe strategy. The International Garden Show was the smaller of the two projects in terms of scope and budget. Led by a corporation that was created by the state for the purpose, the IGS redeveloped 85 hectares of central Wilhelmsburg where there had previously been various land uses, including a city park, allotment gardens, trees, and a container yard. Over the course of roughly

4 Internationale Bauausstellung Hamburg

five years, the area was redeveloped into a sculpted landscape of paths, thematic gardens, grass and water. It culminated in a 2013 show, through which the city aimed to attract visitors to "discover" Wilhelmsburg as a part of Hamburg (Behörde für Stadtentwicklung und Umwelt, 2013). The project claimed to be self-financing based on ticket sales, but as it attracted just 50% of its projected 2.5 million visitors, it ultimately cost the city €37 million (Behörde für Stadtentwicklung und Umwelt, 2013). Much of the foundational landscaping produced for the show remains in the form of the Wilhelmsburg Island Park (*Inselpark*). Figure 28, a photo that I took in the height of summer 2017, shows some of the Island Park today and its location within view of the Ministry of Urban Development and Housing and other IBA Hamburg projects.

Though many of the racialized residents I interviewed in 2017 described the new Island Park as one of their favourite spots in Wilhelmsburg, the IGS that produced the park was a lightning rod. One point of public protest took place in 2010, when it became clear that thousands of trees were being cut down to create the show (cf. Hamburger Abendblatt, 2010). Local activist groups organized to protest this "clear-cut" and to demand accountability from the IGS and the city. The irony of removing thousands of trees to create a Garden Show was not lost on residents, especially in the context of the city's claims to environmental responsibility and sustainability in both the IGS and IBA. Marianne Gross from *Zukunft Elbinsel*[5] questioned why the landscape at the middle of Wilhelmsburg was being reinvented entirely, rather than integrating what existed into the project plans (Hamburger Abendblatt, 2010). Florian Hohenstatt, from the group *Arbeitskreis Umstrukturierung Wilhelmsburg*[6] argued that the goal did not seem to be a natural green space at all; rather the IGS was creating a "respectability theme park" that was "sorted, transparent, and washable" (Hamburger Abendblatt, 2010, translation by author).

Another, related point of protest came later in the IGS process, after its "show" year passed. Though the IGS promised the production of a "high quality" park for Wilhelmsburg (IBA Hamburg, n.d.-a), limitations on access to the park even after it was nominally "handed over to the neighbourhood" rubbed many residents the wrong way. In 2014 the fences that were erected to funnel visitors through ticket booths during the show were still present and local officials were debating whether to maintain them in order to control vandalism. While it was argued by government officials that vandalism was not a problem specific to Wilhelmsburg (David,

5 An activist group that grew out of the Conference for the Future (see Zukunft Elbinsel Wilhelmsburg, 2012).

6 Working Group on the Restructuring of Wilhelmsburg, AKU for short, formed as a platform for critique of the developments on the island. This was a group (disbanded after the 2013 events) of younger and more left-leaning activists than the largely middle-class long-term residents of *Zukunft Elbinsel*.

2015), the message that some residents received was that "this park is too valuable to entrust it to city residents without any limitations" (Trautwein, 2014a, translation by author). Despite "massive criticism" from Wilhelmsburgers (Brück & David, 2014), the fence remains today, and three of the park's entrances are closed nightly from midnight to 5:00 a.m., ostensibly to stop cars from entering the park at night (David, 2015; Zukunft Elbinsel Wilhelmsburg, 2014). One local journalist argues that the fence is a symbol of the divide between the IGS as a tourist attraction and the actual everyday life of the neighbourhood (Lasarzik, 2014).

These critiques of the IGS are shaped by local questioning of who the park is ultimately for. Certainly park-making is not neutral, but rather is influenced by the context and power relations in which it takes place (cf. Barman, 2005; Byrne & Wolch, 2009; Gandy, 2002; Kheraj, 2013). In particular, access to a park – literally, in terms of how and when one can get in, but also symbolically in terms of whether a park feels welcoming and to whom – is constrained by the classed, racialized, and gendered conditions in which it was produced (Byrne & Wolch, 2009). It is significant, then, that the IGS was framed as initially not *for* Wilhelmsburgers, but rather for the discovery of other Hamburgers. The language is distinctly colonial, positioning the IGS as harnessing land to attract interest and potential new arrivals, and Wilhelmsburgers themselves do not appear as the subjects of the development.

The production of the IGS was also marked for some residents by years of lack of access to green space. Gülhan described for example how her family had used the city park that existed before the IGS, and how they had missed it during the many years of closure:

> It was a normal park, but also nice, it wasn't so... now it is obviously a bit more structured. We always used to go there as a family, Saturday, Sunday in the summer, always barbecued, picnicked. Yeah, that was our favourite spot, it is still a favourite spot... same as it was years ago, twenty years ago. It was sad at the time [during the redevelopment], we couldn't go in wherever we wanted. Then it was that we had no access, for a few years (Gülhan, interview, 2017).

This loss of access pushed her family to look for private green space, which they found in the form of an allotment garden. Though allotment gardens are usually collectively owned, they are enclosed spaces over which the renter has control.

Other racialized residents' experiences with the production of the IGS illustrate how parks are often driven by an interest in social control (Byrne & Wolch, 2009). Mohammed told me a story about attending one of the resident-consultation meetings about the IGS, which are required by law in the case of any major developments (Bundesministerium der Justiz und für Verbraucherschutz, 2017; Freie und Hansestadt Hamburg, Bezirk Hamburg-Mitte, n.d.). Though he intended the anecdote to illustrate how resistant people can be to change, and thus to immigrants as perceived bringers of change, it also illuminated some of the power relations embedded in local discussion about the purpose of redesigning the pre-existing park:

> There was an older guy there [who said]: 'The park is so dark, you can't even...', totally mad about various things. 'And who knows *who* is hanging around there' and blah blah blah. But he was also complaining about the new park that would be built... So I, now I have my turn: 'The park as it is now is bad, not enough light, the people who are there, one is afraid to go through, and now someone comes and says, we're making a new park with more open space, lots of light, so that everyone can walk there and so on, and you're also against that. Against *this*, and against *that*: what do you actually want?' He said: 'Oh go back where you came from!' [*laughs*] That was the answer! (Mohammed, interview, 2017)

Mohammed saw in this anecdote peoples' incapacity to contemplate change without feeling threatened and lashing out. What I notice is also the racist attempt to evict him from the process and from authority to speak, and perhaps particularly to ask challenging questions. If racialized Wilhelmsburgers experience such racist hostility in the park-making process, it raises doubts about the accessibility of the eventual park itself. His interlocutor makes a claim that Mohammed is not at home, likely based on his dark hair and eye colour, an experience which as noted in the previous chapter is a very common one for him. This was understandably Mohammed's only foray into the "participatory" planning process, which I will discuss in more detail in relation to the IBA.

What Mohammed's anecdote further illustrates is how the old park was problematized by some residents and not others. The park that Gülhan described as a beloved family spot is characterized here as dark and potentially dangerous. This fear must be understood within the context of discourse about the racialized threat in Wilhelmsburg. The "who knows *who*" raises the spectre of people lurking in the shadows, perhaps particularly the racialized young men who the dominant bourgeois, white society tends to view as threats (Bucerius, 2014; Haritaworn, 2015). The thinning of the trees to which many local residents objected in general is connected to the production of the new park, "the park of the 21st century" (David, 2015), in which there is nowhere left to lurk. This is the "sorted, transparent, and washable" environment that Hohenstatt critiqued; it is an environment determined by a bourgeois urban ideal, which local activists and scholars also identify in relation to the IBA Hamburg (Rinn, 2018).

Figure 28: The rose garden in Island Park, flanked by allotment gardens, with IBA Hamburg projects in the background: the swimming pool, office building, and the Ministry of Urban Development and Housing

(photo by author, 2017)

International Building Exhibition (IBA Hamburg)

In comparison with the International Garden Show, the International Building Exhibition was a much larger project in scale and cost. IBA Hamburg comprised over 60 building projects, €120 million of public investment, and €1 billion of private investment (IBA Hamburg, 2018; Vogelpohl & Buchholz, 2017). This IBA was the latest in a 100-year urban planning and architecture tradition in Germany. IBAs are large-scale projects that create and transform built space, showcasing architecture while intervening in urban issues that are considered to be pressing and otherwise difficult to address (Hellweg, 2010; IBA Hamburg, 2015). Past IBAs have, for example, transformed deindustrialized zones in the Ruhr valley, parts of the former East Germany, and the crumbling housing of 1980s West Berlin (IBA Hamburg, 2015; Lütke Daldrup and Zlonicky, 2010; MacDougall, 2011).

The central goal of this IBA was to make the island of Wilhelmsburg more attractive to the "middle class" (cf. Hellweg, 2013; IBA Hamburg, 2010). To achieve this, IBA Hamburg implemented a wide range of projects organized along three themes: Cosmopolis (harnessing diversity as a strength), Climate (resilience for the low-lying island), and Metrozones (addressing conflict between residential and industrial areas). The projects mainly comprised new builds such as an education centre, a sports hall, and the new home of the Ministry for Urban Development and Housing, which was relocated from downtown Hamburg (see Figures 29 & 30). There were also several major renovation projects, such as the retrofitting of a block of public housing (the Global Neighbourhood, Figures 31 & 32), and the transformation of an anti-aircraft bunker into a plant for "green" energy (Figure 34 & 35). For descriptions and pictures of more IBA projects, see IBA Hamburg (n.d.-b).

One of IBA Hamburg's notable features was that it was framed not only as a development project and an event, but also as knowledge production. The project claimed in its extensive publicity materials, on which it spent an estimated €12.85 million (Bürgerschaft der Freien und Hansestadt Hamburg. 19 Walhperiode, 2009) that IBAs are "urban laboratories" (IBA Hamburg, 2018) and "an experimental field for urban development" (Lütke Daldrup & Zlonicky, 2010, p. 14). On the Elbe Islands, IBA Hamburg would engage in "real-time research and development. Like a laboratory. Except that the laboratory is in fact an entire district in the city and research leads to actual built space" (IBA Hamburg, 2015)[7]. In addition to driving the Leap Across the Elbe, the IBA claimed to address the "major social issues of our time" and to produce blueprints for the future of the metropolis (IBA Hamburg, 2011a, 2015). Some of these blueprints consisted, for example, of "in vivo architectural experiments" (Eckardt, 2017, p. 175), such as the Water Houses pictured inFigure 33. The flood resilient Water Houses are located at the entrance to the Island Park, at the centre of the "bathtub," as the geography of Wilhelmsburg was once described to me. They comprise 34 apartments in total and were built at a cost of €11 million (IBA Hamburg, n.d.-b) and are therefore well outside the housing budget of the average Wilhelmsburg household.

By framing its work as knowledge production, the IBA Hamburg managed tensions and skepticism related to its big ambitions. As architect Frank Eckardt (2017, p. 178) puts it, IBA's expectations appeared "paradoxically high and low at the same time;" it claimed to solve major problems while it downplayed its capacity to change much at all. The website cautioned for example that results of the experimentation would not be known for some time, not until after the 2013 exhibition year, "once everyday life resumes on Hamburg's Elbe Islands" (IBA Hamburg, 2018). By projecting ambition and caution simultaneously, IBA Hamburg mobilized an image of a hopeful ("better") future Wilhelmsburg while also holding off questions about how, precisely, one architectural exhibition would solve the major issues of our time (Eckardt, 2017).

7 See Chamberlain, 2020 for my analysis of the use of urban laboratory discourse in the context of racialized space.

I have argued elsewhere based on my master's research, that IBA Hamburg contributed to the racialized stigmatization of Wilhelmsburg and its residents in its extensive publications and marketing materials (Chamberlain, 2012, 2013). The IBA's materials reproduced the characterization of Wilhelmsburg residents as problematic and in need of intervention, while also obscuring the ongoing presence and everyday lives of tens of thousands of people who lived on the island before, during, and after the mega-project. Except for where it framed diversity as a consumable selling-feature, the IBA's focus was squarely on incoming, middle-class residents and not on the lives of existing Wilhelmsburgers. As the following chapter will discuss, some of my research participants thus analyzed development on the island as oriented towards the needs and desires of "new" rather than "old" Wilhelmsburgers.

The planning strategies implemented by the city-state of Hamburg in Wilhelmsburg – subsidized student housing, social mix in public housing, and the IGS and IBA event-projects – attracted substantial critique in scholarship and activism based on the island. There is a substantial body of research about development in Wilhelmsburg, and particularly about IBA Hamburg on the island. I draw mainly on local analysis of the city's interventions as state-led gentrification, and as the production of specifically bourgeois space. While the latter was transparently intentional on the part of the city – it was explicitly embedded in the logic of the interventions, as I have illustrated so far – the former is contested based on varying definitions of gentrification and varying assessments of the unfolding facts in Wilhelmsburg. The following section explores this contestation, before concluding with my own analysis of how the racialization of displaceability and thus the risk of un-homing is under-emphasized in gentrification critiques.

Figure 29: A view of IBA Hamburg construction in 2011, where the new Ministry of Urban Development and Housing building went up. Looking north-east from Neuenfelderstrasse, with the mall and some high rises around Berta-Kröger-Platz in the background

(photo by author, 2011)

Figure 30: The completed IBA project 5 years later, looking westward at the added footbridge connecting to the transit station and mall

(photo by author, 2016)

Figure 31: The façade of part of the renovated 'Global Neighbourhood,' at the corner of Veringstrasse and Neuenfelderstrasse

(photo by author, 2016)

Figure 32: View westward from the bunker, with the Global Neighbourhood IBA project on the left side, and unrenovated housing of a similar age across the street

(photo by author, 2017)

Figure 33: The Water Houses at Island Park, part of the 'Legoland' built by IBA Hamburg in central Wilhelmsburg

(photo by author, 2016)

Figure 34: Views of the bunker overlooking the Haus der Jugend Wilhelmsburg in 2011 (34) and in 2017 (35) after it was turned into a power plant.

Figure 35: The bunker protected thousands during WWII, though not the racialized enslaved workers who built it. Like other bunkers still standing in the city, it proved too difficult and expensive for the British to remove after the war (IBA Hamburg & Geschichtswerkstatt Wilhelmsburg und Hafen, 2013; Rossig, 2014). Today it has a fancy café

(photos by author, 2011, 2016)

Figure 36: Construction has continued since the IBA, adding the 'Woodie' student housing

(photo by author, 2017)

Gentrification, bourgeois space, and the threat of displacement

Hamburg's urban planning strategies in Wilhelmsburg, particularly the IBA and IGS, were explicitly aimed at the production of space for more affluent people, which is one of the broad definitions of gentrification (Hackworth, 2002). Indeed, the three prongs of the city-state's interventions have all been identified elsewhere as faces and forms of gentrification. The student subsidy raised alarm bells in local media (cf. Kleinhubbert, 2008; "Subventioniertes Wohnen," 2013), as students often play a role in "marginal gentrification," in which people who do not have particularly high incomes act as the cultural vanguard for the eventual arrival of people who do (Elliott-Cooper et al., 2020). The encouragement of students to take over apartments that could house families, in particular is an aspect of what has been called "studentification" (D. P. Smith, 2004), which is linked to retail gentrification and exclusionary displacement, wherein people lose access to the kinds of units that they would have previously been able to rent (Hohenstatt & Rinn, 2013, drawing on Marcuse, 1985). New-build development, which comprised the majority of IBA Hamburg, is similarly likely to produce displacement slowly, over time, by "making working-class residence ... increasingly untenable" (Elliott-Cooper et al., 2020, p. 6). Local researcher Moritz Rinn thus characterizes Hamburg's planning as an "urbanism of inequality" that uses displacement as a planning tool (Rinn, 2018, translation by author).

Hamburg's planning strategies in Wilhelmsburg hinge to an extent on the encouragement of what are sometimes called "urban pioneers" (a troubling term with colonial connotations that typically go unmentioned; see Christmann for an example, and Haritaworn, 2015 for critique). Urban pioneers – artists, social entrepreneurs, the self-employed, and in Wilhelmsburg's case perhaps also students – appreciate the aesthetics and opportunities that run-down and devalued neighbourhoods offer them (Christmann, 2013, 2014). They often push against the dominant narratives about stigmatized neighbourhoods, articulating a different valuation of space (Zwilling, 2012), while driving transformation out of their own imagination and desire to use that space (Christmann, 2013). One study of Wilhelmsburg argues that the IBA itself functioned as an urban pioneer by bringing "new" and "creative" energy to the long-neglected island. It eventually became "part of the problem," however, as it solidified images of difference and exclusion (Zwilling, 2012, translation by author), as I also argued, albeit in different terms (Chamberlain, 2012, 2013).

One of the major local critiques of the IBA in particular was that the resident participation experience was more about producing acceptance than actual participation. Some of the people and organizations I encountered in Wilhelmsburg had participated actively in early phases of the planning, as far back as the Future Conference, but had eventually tapered off their participation and tempered their initial enthusiasm as a result of discouraging and alienating experiences with the project (see also Schmidt, 2012; Zwilling, 2012). Manuel Humburg, a long-time resident, organizer, and recently retired neighbourhood doctor, argued that the resident engagement processes in IBA and other projects undermine local resistance and minimize conflicts by diffusing and burning out people's energy (Manuel, interview, 2017). The limits and contradictions of development in Wilhelmsburg were largely externally determined, including by the port, as I mentioned in Chapter 3, and by the Hamburg senate, which signed off on a new coal power plant and the dredging of the Elbe river just as the IBA was supposed to be searching for strategies to protect against climate change and flooding (Humburg, 2009). Further, the IBA shielded the senate and parliament from accountability and resident protest; criticisms of the IBA were often fundamentally political questions for which the IBA as project management was the "wrong addressee," as a planner put it to me (anonymous interview, 2018).

This is not only the case with IBA Hamburg, so it is worth detouring briefly into the present to note that the way the terms of development are already set by political decisions that take place well before consultations begin is at the core of residents' frustrations with participation in general. As Heike Sudmann (*Die Linke*) put it for example, "there are a lot of offers of participation, I would say cautiously ... but it is not the case that you can really change anything, because it is usually predetermined... This is naturally a participation that frustrates" (Sudmann, interview, 2018). I found that many of the residents I interviewed felt alienated from a political process that had made the decisions beforehand. Their analysis was that politicians do not listen to the average Wilhelmsburger, but rather attend to the interests of people with property, money, and power. Umut argued for example that politicians and property owners are closely tied, and "that's why they just have more pull, because they have close contact to politics" (Umut, interview, 2017). Gizem felt similarly and mused about a possible future in which her children and other Bulgarians' kids were grown up and had entered positions of power and influence. Then, she argued, it will not be possible for politicians to forget about us (Gizem, interview, 2017).

Other Wilhelmsburgers argued that they simply do not have political representation because that is structured by German citizenship, which a third of Wilhelmsburgers do not have. "We are not allowed to vote in Hamburg, so not in the provincial parliament, and not at the federal level either. That is a very, very powerful racist rule in this country. That is the most blatant example that I can give you. Voting rights have been demanded by migrant self-organization for 40 years, yes, and that has done nothing" (Ali, interview, 2017). Metin Hakverdi (SPD), who represents Wilhelmsburg in German parliament, also recognized that this is an issue at the national level in Berlin. While he suggested that citizenship should be an instrument of "integration," rather than a reward at the end of a process, he also noted that those who do not want to allow dual citizenship get a lot of support in federal elections. Hakverdi questioned, as well, whether the level of current electoral participation among people who *do* have the right to vote really suggests that it is an urgent issue. Yet some Wilhelmsburgers did argue that it is an important issue both materially, in terms of the possibility of influencing political decisions with one's vote, and symbolically in terms of the visibility of racialized people in politics. Pastor Obeng again compared Germany to England, as a reference point with which he is familiar. In Britain, he argued, the children and grandchildren of immigrants are visible in key political positions. In Germany, in contrast, the government is "confused" as he put it. "Politicians need to update their thinking, that is all I can say" (Peter, interview, 2017).

When I spoke to Sören Schaefer, who works for a project called *Perspektiven! Miteinander planen für die Elbinsel* (Perspectives! Planning together for the Elbe Islands), he argued that a key role of participation in local development is to bring people into democratic processes. Perspektiven! plans and facilitates Wilhelmsburgers' participation in all of the current development projects on the island, including those for which IBA Hamburg now functions as a project developer (Bürgerhaus Wilhelmsburg, 2018; IBA Hamburg, 2019). Schaefer noted that "in Wilhelmsburg it is the case that there is a large number of people who are shut out of democratic opportunities, because they don't have the passport" (Schaefer, interview, 2018). Democracy in that sense does not extend to everyone, and in any case suffers legitimacy issues wherein fewer and fewer people choose to participate in elections. Perspektiven! imagines planning participation as a democratic opportunity in this context: "We believe that one, that everyone can say something about the neighbourhood, about the direct environment" (Schaefer, interview, 2018).

When I was conducting this study in Wilhelmsburg, I participated in the consultation process that was underway for the Spreehafenviertel, a 20-hectare triangle bounded by the Ernst-August Canal to the south, Schlenzigstrasse to the east, and the Spreehafen to the north (IBA Hamburg, n.d.-c). I lived directly across from this area, as I have shown in photos throughout the book. Presently the 20 hectares is comprised of a variety of land uses, including one of the longer-term refugee housing compounds and the alluvial forest that I mentioned in Chapter 3. The consultations included public meetings, an online survey, and various information stands.

Among my observations of the consultations was that the attendees at the public meetings were overwhelmingly older white people, and the meetings took place entirely in German at a speed that I struggled to follow. By no stretch of the imagination can they be considered accessible or inclusive to a diverse, multi-lingual community. This is something that Perspektiven! recognizes: "We live in a hyperdiverse neighbourhood, so we absolutely have to ensure, or at least make an effort to reach people ... we do that a lot, where we do interviews, so go to where the people are" (Schaefer, interview, 2018). In this way, Perspektiven! aims to reach people from different "social milieus" to participate in a range of ways. They reject the notion of "migration background" on the grounds that it is divisive and exclusionary. "For us, other questions are more relevant than [whether or not] someone's grandparents at some point immigrated.... In social milieus there are many more intersections than just culture of origin" (Schaefer, interview, 2018).

The questions put to residents in consultations on the *Spreehafenviertel* were indeed not about whether the development should happen at all, but about how it should be laid out. Alongside the process, resident protest emerged and organized under the banner "*Waldretter*" ("forest savers"). The *Waldretter* call for a halt to the *Spreehafenviertel* development and a halt to development in Wilhelmsburg altogether. The group's central message is that the Spreewald should be protected as an important habitat for wildlife and an ecological resource for a neighbourhood that carries a heavy but poorly documented environmental burden. Further, they argue that no new development should be built on green spaces in Wilhelmsburg, and that planning requires an overall pause to assess the environmental impact of the various projects already in the works (Waldretter, 2019). As there is no possibility *within* the consultation process for such resistance to find an interested audience, they demonstrated outside the meetings, and hosted a separate forum for heated community debate. The group's demand to stop building entirely contradicts what I heard from racialized residents about the urgent need for housing on the island, which I will explore in the following chapter. The *Waldretter's* position is that what is built in the *Spreehafenviertel* will not be affordable anyways. The demand for protection of trees and green space dovetails with the crucial importance that long-time residents placed on the greenness of the island, as I discussed in the previous chapter.

At the time when the IBA and the IGS were underway as event-projects, local activism forced Hamburg planning and the IBA itself to respond to the critique of state-led gentrification. Vocal activism emerged particularly from *Arbeitskreis Umstrukturierung* (AKU) Wilhelmsburg, which had crystallized around the most visible aspects of the city's interventions, the IBA and IGS. IBA Hamburg, and the Leap Across the Elbe more broadly, claimed to follow a program of "improvement without displacement" (*Aufwertung ohne Verdrängung*), a motto that in fact emerged around 2010 as a response to resident activism (Vogelpohl & Buchholz, 2017). At the centre of this claim was the assertion that the IBA would not put pressure on the living situations of current residents, because it mainly constructed new housing on land that was previously non-residential (cf. IBA Hamburg, 2013). I also heard this perspective from politicians such as Metin Hakverdi (SPD), who presently sits in the Bundestag for Wilhelmsburg. He argued that the island had been in a unique and lucky position in this regard: it had proved possible to change the composition of the population, bringing in more people with higher incomes without throwing others out.

Critical local voices challenged the claim that displacement would not be a factor in Wilhelmsburg, however, arguing both that IBA Hamburg *had* directly displaced some residents and that it was likely to produce further indirect displacement over time (Arbeitskreis Umstrukturierung Wilhelmsburg, 2011, 2013a; Hohenstatt & Rinn, 2013). The debate converged particularly around the so-called "Global Neighbourhood" (Eckardt, 2017), the one IBA project that involved the renovation of social housing. The estate, owned by Hamburg's public housing company SAGA GWG, comprised several blocks of apartments that were built in 1930 for workers at the Howaldt shipyards. The IBA dubbed the estate the Global Neighbourhood to emphasize the diversity of the people who lived there as both a problem and an attraction (see Chamberlain, 2012, 2013); in 2007 there were about 1,700 residents "from 30 different countries" living in 823 apartments (SUPERURBAN PR und Stadt planbar, 2007, p. 4).

The project carried a substantial burden of meaning for the IBA, for the city, and for gentrification researchers. The project, in which IBA Hamburg worked together with SAGA GWG to renovate and restructure apartments and shared spaces, promised the "social stabilization" of the area and a "model project for intercultural living" (Hellweg & Basse, in SUPERURBAN PR und Stadt planbar, 2007, p. 3, translation by author). Part and parcel of the IBA's framing of the project is the invisibilization of existing residents through such claims; the IBA's narrative tended to devalue residents' existing lifestyles and position them as waste to be cleaned up in order to appeal to middle-class tastes and values (Eckardt, 2017). This reflects the ongoing pattern of devaluation of racialized, low-income people that I identified in the previous chapters.

The Global Neighbourhood played a central role in IBA's self-legitimation, though it was in fact unusual among its 60+ projects. The IBA Hamburg claimed in general that "no major decisions [were] to be made without the support of the people living on the Islands" (IBA Hamburg, n.d.), and the source of support and consent that was routinely referenced was a survey conducted in the Global Neighbourhood before it was renovated. During my own research, this survey was regularly mentioned to me as the evidence base for the IBA and as something that would interest me because multilingual students went door-to-door to talk to residents about prospective developments. The "*Heimat* researchers," as they were called (*Heimatforscher*), started conversations with residents who were then invited to visioning workshops (SUPERURBAN PR und Stadt planbar, 2007). While this may indeed have been an exemplary engagement process in a context where it appears unusual to consult people in languages other than German, it cannot be considered comprehensive research that validates the whole of the IBA. It spoke to the residents of one city block about the renovation of that block. The survey itself was impressionistic, as it asked only "*Was ist Heimat für Sie?*" ("what is *Heimat* for you?), a most interesting and poignant question given *Heimat's* fraught meanings, its frequent use to exclude racialized and migrantized people, and Wilhelmsburgers' assertions of the island as *Heimat* and thus as a place where they wish to stay (see the previous chapter).

When it came to the Global Neighbourhood, resident activism and scrutiny of the housing situation in Hamburg encouraged the IBA to "[reintroduce] a minimum of respect for vulnerable residents" into housing (re)development (Vogelpohl & Buchholz, 2017, p. 273). Though the mandatory commitment of particular apartments to social housing had been dropped in 2003 to 15 years as part of the CDU's neoliberalization strategy, the IBA negotiated for a return to the previous 30-year status in the renovation of the Global Neighbourhood, and guaranteed residents the right to return at a fixed rent (Vogelpohl & Buchholz, 2017). Debate persisted, however, about how many residents actually returned to the estate and whether it should be considered acceptable that rents for the new apartments were higher per square metre and the apartments larger (Arbeitskreis Umstrukturierung Wilhelmsburg, 2013b; Eckardt, 2017; IBA Hamburg, 2013; Sassen, 2013).

The differences of opinion on the Global Neighbourhood were based on different assessments of the facts, but also on different meaning made of those facts. Local activism and scrutiny by AKU and others linked to Hamburg's vibrant Right to the City movement (see Recht auf Stadt – Netzwerk Hamburg, n.d.) pushed the IBA and the city-state to take the threat of gentrification seriously, and this led to the implementation of a temporary "structural monitoring" process to track whether there were signs of gentrification (cf. Analyse & Konzepte, 2013), and to an info sheet on gentrification that can be found among the IBA's many publications (IBA Hamburg, 2013). These documents reflect the official position that gentrification was not taking place in Wilhelmsburg because 1) there was no evidence that anyone had been directly displaced, 2) rents, though rising, were still below the Hamburg average, and 3) there was no noticeable luxury renovation or buying up of rental housing (IBA Hamburg, 2013). Further, they argued that the "*Drittelmix*" (mix-of-thirds) formula protects against gentrification on the island in general.

There is fundamental disagreement about whether these measures reflect sufficient protection of low-income people in the long run. Hamburg has an affordable housing shortage (cf. Neues Deutschland, 2014), and around 40% of the city's entire population is entitled to subsidized housing based on their income (Gaßdorf, 2018; Vogelpohl & Buchholz, 2017). The status of much social housing also continues to expire, resulting in a city-wide environment of shrinking access in the face of growing need (Twickel, 2017; Wenig, 2018). In this context the IBA's elision of clearly visible signs of gentrification, which I will turn to in the following chapter, and its argument that things are not as bad in Wilhelmsburg as elsewhere in the city, is more frustrating than placating to local residents and activists.

At the core of gentrification is a process of displacement, in which lower-income people are "un-homed" (Elliott-Cooper et al., 2020). That process can take a variety of forms and temporalities: it can happen quickly and directly, and indirectly over time (Elliott-Cooper et al., 2020); if it involves the replacement of lower income residents by more affluent people, then it can be correctly identified as gentrification (Bridge et al., 2012; Hackworth, 2002; Slater, 2006). However, the lens of gentrification has not helped to illuminate the specifically racialized displaceability that is produced by the city-state's interventions on the island and that is intertwined with Hamburg's long history of producing and reproducing some (racialized people) as objects rather than subjects of urban planning. Racialized threat has in fact contributed to the very basis of interventions, as I discussed in previous chapters, and which is also clear in the concern with "too many foreigners" as it is with the wild disorder of park space. Local scholars note that Hamburg's planning involves governing through inequality (Rinn, 2018), and that the governance uses control and exclusion (social mix policy) along with market incentives (student subsidies) and attraction through neighbourhood improvement (IBA and IGS) as a form of population politics along the logic of social mix (Hohenstatt & Rinn, 2013). This ensemble of planning institutions and practices aims to produce and manage space based on bourgeois norms and ideals (Hohenstatt & Rinn, 2013; Rinn, 2018), but these ideals are specifically white and German.

The belief that social mixing will "strengthen" and "stabilize" communities, as is articulated in the policies I have quoted above, depends upon the production of racialized people as fundamentally different, non-German, and therefore *de*stabilizing.[8] This kind of logic, which devalues and problematizes based on race embodies the legacy of colonial thinking that scholars and activists have identified in Hamburg. The social mix policy in public housing illuminates the role of race in the city's calculus in Wilhelmsburg most starkly. As discussed above, public housing providers pass over applications from the lowest income people and from people "with migrant-sounding names" *regardless of income* (Adanalı, 2013). This practice depends upon the assumption that you can tell if a person is a migrant based on their name, and that "migrants" are fundamentally distinct and problematic to the community. The terms in which social monitoring is undertaken in Hamburg underscore this assumption; the notion of "too many" as an urban planning challenge is embedded in Hamburg's problem definition, data collection, and problem-solving strategies. Insofar as to govern is to know (Rodatz, 2012, drawing on Foucault), Hamburg seems mainly to want to know who lives where and how many. I asked every politician and planner I met in Wilhelmsburg how the city-state would know if its goals were met on the island

8 See Chamberlain et al. (2022) for a roundtable discussion of racism and social mix from an international perspective.

and the answer I consistently received was "good question!" Social monitoring and pre-existing forms of demographic data collection, such as the micro census and resident registration, are the only forms of evaluation that the city has planned. I also noticed signs that the city itself realized that social mix was a faith-based policy. At a planning consultation in Wilhelmsburg in 2018 I overheard an official from the district level say that the city is "convinced" that social mixing is the way forward, *although they know that the evidence might not be sound*. "We are convinced nonetheless," the official asserted (paraphrase of personal communication, 2018).

As interventions geared towards "mixing" Wilhelmsburg, the city-state's strategies on the island thus indeed appear as a "racialized technique of recasting relations of domination" (Kipfer & Goonewardena, 2014, p. 103). As Adanalı puts it, "the city created the 'social flashpoint' of Wilhelmsburg itself, so to speak. After migrants settled in Wilhelmsburg and built up their social environment etc., they now have less chance to live in their neighbourhood as a result of social mixing" (Adanalı, 2013, p. 123, translation by author). Racialized Wilhelmsburgers thus appear to be the objects rather than subjects of (re)development, as their ability to claim home and land, as Roy (2018) puts it, is curtailed in favour of mix and thus the access, priorities, and well-being of white, middle-class Germans.

This is facilitated by the expression of white supremacy in German law, based on geographer Laura Pulido's (2015) definition of the concept. In an exploration of a specific example of the state's role in urban environmental racism in the United States, Pulido identifies how state practices reflect a belief that white people are entitled to more than racialized people because they are believed to be "more deserving, or 'better' than other groups" (Pulido, 2015, p. 812). Pulido argues that though the concept of white supremacy is poorly understood (associated primarily with extremists and Nazis, for example) its circulation is visible in "fears of a non-white nation, whites' sense of ownership and the right to exclude, and deeply racialized thinking that systematically undermines the well-being of people of color" (Pulido, 2015, p. 812 drawing on Rattan & Eberhardt, 2010).

Pulido outlines three aspects of white supremacy whose presence I am struck by in the implementation of social mixing in housing in Wilhelmsburg in particular: awareness, taking, and belief in racial superiority (Pulido, 2015). An awareness that racialized people are being harmed, in this case through the restriction of access to affordable, public housing, might perhaps be obscured by devotion to the ideology of social mix on the part of the city-state of Hamburg. However, the city-state might reasonably be expected to know that this exclusion creates hardship, and certainly scholarly evidence of the harms associated with a lack of affordable housing and lack of choice in housing is abundantly available to the city and to urban planners. Further, the very existence of the General Law on Equal Treatment demonstrates awareness on the part of Germany as a whole that discrimination causes harm that needs to be avoided. Yet the city-state takes a resource from non-white people, here in the form of access to a fundamental public good: affordable housing. The belief in racial superiority underpins both the racializing concept of "migration background," and the assumption that ghettoization and destabilization of the neighbourhood are caused by the presence of too many racialized people. This belief is authorized and encouraged by the federal state through its production of the *Migrationshintergrund* statistical category and the right on the part of public housing providers to discriminate on this basis. The city-state of Hamburg operationalizes this belief through the actual practice of mixing and through the ongoing measurement of problems and solutions through the counting of bodies.

The practice of white supremacy that is evident here is consistent with the functioning of racial capitalism as it depends upon the devaluation of the lives and well-being of racialized people. Thus, while Adanalı (2013) and Münch (2014) call for explication of criteria for social mixing and transparency in decision-making and for oversight to ensure that discrimination is not taking place, I find that the practice is discriminatory by nature and a process of devaluation in and of itself.[9] Wilhelmsburg might benefit from an open and frank address of what precisely social mixing means in practice, considering how the politicians I spoke to voiced support for mixing along class lines, but not according to "ethnic" or "national" criteria. This suggests to me that local actors are either not fully aware of how social mixing is implemented in Wilhelmsburg – the housing policy is likely not publicly advertised and marketed for a reason – or that they look away from the details in the interest of developments that are otherwise seen as much needed.

Mixing presents a real problem for racialized people who are deeply attached to the island as *Heimat*. The city-state is not invested in the cultivation of this *Heimat*, but rather in dispersal and discouragement of further concentration of racialized people in Wilhelmsburg. Though many residents are comfortable with the notion of mixing, insofar as it promises to counter the isolation and stigmatization that they have experienced for decades, as I will explore in the following chapter, they are also concerned about the well-being of the most vulnerable among them. They see evidence that migrants and racialized people are being pushed to move elsewhere or are getting stuck in poor living situations for lack of other options. Both of these are forms of displacement (Slater, 2006), and thus of state violence (Roy, 2018) to which racialized people are deliberately – and legally – made vulnerable. Yet the city-state's interventions are largely seen as welcome, if imperfect, from the hopeful perspectives of long-time residents.

9 For further discussion see Chamberlain et al., 2022.

Conclusion

In this chapter, I have presented the city-state of Hamburg's main planning interventions in Wilhelmsburg as they were identified to me by research participants and characterized mainly in government archival documents. I outlined the subsidy for student housing, social-mix policy in public housing, and twin event-projects that Hamburg has implemented since 2005, arguing that they are based on the planning common sense of social mix. I have situated the planning strategies as contested and critiqued from the island and beyond, drawing mainly on local scholarship and activism and my own interviews to highlight some of the points of contention. The event-projects in particular have been identified as tools of state-led gentrification in Wilhelmsburg, aimed at producing bourgeois space despite the claim to improvement without displacement. I argued that gentrification does not capture the racialized production of displaceability in this case, and that the threat of un-homing specifically of racialized and low-income people needs to be situated as a function of racial capitalism enabled by white supremacy in German law. The racialized production of displacebility can also be understood as part of Hamburg's colonial legacy, as I discussed in Chapter 3; in the city's planning logic, some people are more displaceable because they are understood as worth less than others.

In the following chapter, I turn to Wilhelmsburgers' analyses of whether and how the city-state's interventions have changed anything on the island, and whether they promise to in the future. I expected to encounter skepticism or even hostility towards the notion of mix and the related arrival of "Germans" to the island, the main development that racialized residents reported thus far. What I found instead was hope that mixing would end Wilhelmsburg's stigmatization and with it the marginalization of its residents. In the context of racism and exclusion, residents are committed to conviviality, and hope for more connection across difference and for new neighbours who will engage with them. Yet neighbourhood change, to the extent it exists so far, is not without conflict; the needs and desires of "new" and "old" Wilhelmsburgers are sometimes different and opposing.

6. Mixed Feelings about Neighbourhood Change

Introduction

Every time I have returned to Wilhelmsburg I have seen something new. In 2019 there was another new café in Reiherstieg and major construction on the dike, part of the program to raise the level of flood protection by 80 cm that year (Landesbetrieb Straßen, Brücken und Gewässer, n.d.). In 2018, when I was on the island for much longer, I noticed new cafés and restaurants in Reiherstieg and Wilhelmsburg centre, student residences that had not been there the year before, as well as a yoga studio, a knitting shop, and a vegan grocery store. One of the changes I found most startling was the disappearance of the refugee reception centre on Dratelnstrasse. When I happened upon the nearly empty expanse where it used to be, I was taken aback. Perhaps I had not fully understood the temporariness of the structure, yet it seemed strange that something of such scale could be entirely gone, with just a few eerie traces left behind (see Figures 40 - 42). I wondered where the people who had lived there had gone.

My research questions for this study did not explicitly assume that change would take place in Wilhelmsburg as a result of the city-state's recent policy and development interventions, and indeed this chapter will demonstrate that residents' perspectives differ on whether much, or anything of substance has changed in Wilhelmsburg recently. Yet to an extent an assumption of change was embedded in my semi-structured interviews with racialized residents. I found that asking about whether and how long-time residents had observed recent change in the neighbourhood offered a way into residents' perceptions and analyses of development that did not demand a prior notion of urban development or planning, nor an awareness of specific city-state interventions. I asked broad questions that I could follow up using my interviewees' own terms, questions about neighbourhood change, challenges, and strengths, and about their visions for the island's future.

In this final chapter I draw mainly on my interviews with residents to explore their perspectives on recent developments in Wilhelmsburg, however they framed them, and put their perspectives in conversation with my participant sensing and media archival research. The changes that I heard about most frequently from residents involved the arrival of new people to Wilhelmsburg and an increased "mix" in the neighbourhood. While Wilhelmsburgers were influenced by the common sense that I explored in the previous chapter, they also viewed "mix" from a somewhat different angle. Wilhelmsburgers tied mix to what they see as positive and negative developments in the island's landscape, ranging from welcome new cafés and shops to problematically scarce housing and increased housing prices that threaten the well-being and life chances of the neighbourhood's most vulnerable. Residents understand social mix in some counter-hegemonic ways in this context, critiquing it as a discourse that distracts from housing discrimination, and holding hope that it can challenge white Germans' tendencies to flee and self-segregate.

I explore residents' observations and analyses thematically. In the first section, I touch on residents' assessments of what constitutes "change" for the neighbourhood and what does not. The biggest change is in population, which residents discussed in relation to two key figures: "Germans" (a short hand for white, middle-class Germans constructed in distinction to racialized long-time residents) and "Bulgarians" (a short hand for working-class Bulgarians, Romanians, and other Eastern Europeans who have come to Wilhelmsburg since these countries joined the European Union). I use quotation marks throughout the chapter to denote where I am discussing what I understand as *figures* of "Germans" or "Bulgarians," categories that are generalized to the point of abstraction, as opposed to specific people who are specifically Bulgarian or German. Residents link the arrival of white Germans to the changing landscape of cafés, restaurants, and shops in some parts of the island, tracing the changes back directly and indirectly to the city's recent policies and projects. The arrival of "Bulgarians," in contrast, is understood by some residents as no change at all, but rather a continuation of the island's long history as an immigrant neighbourhood.

In the second section, I focus on the theme of housing, which was on the lips of every Wilhelmsburger I interviewed. The housing situation has changed on the island in recent years, becoming more expensive, unavailable, and inaccessible particularly to racialized people. The results can be quite devastating, as housing is a key to life and life chances in Germany. Yet as I explore in the third section, Wilhelmsburgers are conflicted about the links between housing exclusion and the notion of "social mix" that they generally support. Residents view the "mixing" that is currently happening as a measure to undo the long-standing stigmatization and isolation of Wilhelmsburg, as the previous chapters detailed, and are hopeful that mix will increase contact with white Germans and include them in the convivial landscape that I illustrated in Chapter 4. Wilhelmsburgers also recognize, however, that "mix" does not necessarily mean an absence of conflict. In the final section, I explore some of the space-related tensions that emerged from my interviews, participant sensing, and media archives. Residents described problematic attitudes and tensions in relation to Eastern Europeans' perceived uses of public space, and discourses of fear and deviant use of space contribute to the construction of "Bulgarians" as still separate and distinct from other racialized long-time Wilhelmsburgers. I thus explore differences between and among racialized people from various angles, and again illuminate differential racialization. There is also distance and conflict between "Old" and "New" Wilhelmsburgers, categories that some Wilhelmsburgers use to denote race and class in relation to white, middle-class Germans (i.e. New Wilhelmsburgers). It remains to be seen whether the hopes that residents hold out for mixing will or can be realized in the context of differential power and resources. For many of those I spoke to, affordable housing that is accessible to racialized people is the number one need, but as I discussed in the previous chapter and will return to in the Conclusion, the city-state's strategies of development and the notion of mix itself depend upon the production and exploitation of racialized displaceability, and thus are stacked against this hope for the future.

The changing face of Wilhelmsburg

Throughout this book I have emphasized the vastness, varied landscape, and diversity of Wilhelmsburg, in order to illustrate some of the complexity and richness that residents described to me, and that has historically been obscured by the devaluation of the island. In an effort to impress upon me just how varied the island is, a local organizer for example went so far as to say that "there is no such thing as 'Wilhelmsburg'" (personal communication, 2016), which is to say that the island defies a singular definition or analysis, especially of recent interventions. My interviews with racialized residents bear this assertion out, to an extent, as residents' experiences of the island in recent years were varied, geographically specific, and at times conflictual.

On the one hand, I heard from some Wilhelmsburg residents that little had changed on the island lately. People who lived in the Bahnhofsviertel and Kirchdorf-Süd felt that development had as yet passed them by. The Bahnhofsviertel was characterized as mostly "forgotten" (Gülhan, interview, 2017), marked by neglect and living conditions that are determined by the willingness of landlords to renovate outdated buildings. Mostafa (interview, 2017) suggested that this is a function of the private and for-profit ownership of most of this area, which the city-state has been reluctant to subsidize with revitalization initiatives.

Unless they lived in former IBA projects, as a few of the people I interviewed did[1], residents tended to see the IBA and IGS developments as largely superficial additions to the landscape. The structural disadvantages that I have described and analyzed in the previous chapters continue to be pressing problems from residents' perspectives. Murat does work that involves home visits across the neighbourhood, for example, and he argued that "there is great poverty here... I can tell you that child poverty, elder poverty, and poverty in general is a big problem" (Murat, interview, 2017). Residents were concerned about persistent issues like limited recreational opportunities for young people and having to leave the island to access specialist healthcare, which can be a struggle for Wilhelmsburgers who live on low or fixed incomes and thus need to manage every €3.50[2] they spend on public transit (one way). With concerns like these in mind, residents like Gizem (interview, 2017), told me that though they had noticed new buildings and the new Island Park, "that doesn't change our lives. It's a nice place, but our lives don't change."

Yet at the same time, Wilhelmsburgers often noted that changes they saw as significant *had* been taking place recently, in the arrival of more "Germans" to the island, which is to say *white* Germans, i.e. Germans who are not constructed as having a migration background. Racialized Wilhelmsburgers tended to simply refer to them as Germans which, as I have argued in the Introduction and in Chapter 4, itself illustrated the whiteness of the category and the way in which that whiteness constrains the use of language. Sometimes residents tied this arrival directly to the city-state's interventions on the island, as Casim did for example.

> Since this IBA, this campaign, after that a lot of students moved here, and you see it too on Veringstrasse by the market square, when you go for a walk, you see a lot sitting ... it's good. Lots of German students, lots of Germans are moving here. Wilhelmsburg's reputation was not so good, but for the last few years, you can really see ... it is a good change for Wilhelmsburg (Casim, interview, 2017).

The area that Casim describes, the Reiherstieg or Old Wilhelmsburg part of the island, was an area where subsidies were offered for many years to students signing new rental agreements, through a program which continues today in other parts of Wilhelmsburg, as I have described in the previous chapter. While only a few of the residents I spoke to mentioned the subsidy itself, they seemed to observe the results quite clearly: a swelling of the student population in Wilhelmsburg.

Along with students, residents perceived an influx of young, mostly white German professionals and artists. As Paula put it:

---

1 Two of the residents I interviewed lived in the Water Houses and New Hamburger Terraces respectively, and one had parents living in the multicultural senior living project (see IBA Hamburg, 2016).

2 As of May 2022.

> I think what one notices is just that a lot – or not a lot, but some – Germans are moving here, who are a bit alternative, a bit green-oriented. You really see that a lot in old Wilhelmsburg, so Fährstrasse, Veringstrasse. A lot of nice things have developed as a result, some shops and cafés and such... In Moorwerder, I have a neighbour who also takes the bus, she moved here from St. Pauli a year ago, and she likes Moorwerder and Wilhelmsburg so much that she wants to buy a house. That happens too. They come from an 'in' neighbourhood that is bustling with life and they like it here. It is also young people, mid-30s, who then also want to have kids here. So, that has changed things a bit (Paula, interview, 2017).

Like Casim and indeed all of my interviewees, Paula locates Reiherstieg as the part of the island where she has observed the biggest changes in population. Yet Moorwerder, the more sparsely populated opposite end of Wilhelmsburg, is also in the picture, and a place to which young, white Germans have been drawn from the other side of the river.

In Reiherstieg in particular, "a scene has developed" (Mostafa, interview, 2017) with cafés, restaurants, and shops that cater largely to the new arrivals, as well as to long-time residents who can afford it. Racialized residents saw this as an overwhelmingly positive development. "It used to be that everywhere was Turkish cafés, which I don't go to" said Casim, "now there is the pizza place, hamburger place, and kumpir [baked potatoes], and this Portuguese place, I like it a lot." Some restaurants have been around longer than others; the Portuguese and Italian places were there when I first visited in 2011 (the latter the creation of sisters whose parents immigrated as "guest workers"), while others have opened since 2016, or even more recently. Nalan also mentioned a local brewery that had just gotten started when we spoke in 2017, as well as a local sustainable agriculture initiative. Infrastructure like major grocery store chains have also changed the landscape: "They built a nice Edeka, new, and they built a new Lidl. That had also been missing in Wilhelmsburg, of course, there wasn't... You go in and each has its own customers. You see it too, if you go in [*comic dramatically*] 'Where did *they* all come from!? How are there so many *Germans* living here!?!'" (Casim, interview, 2017).

Wilhelmsburgers linked these developments in a general sense to the city-state interventions of which they were aware. Member of Hamburg Parliament Michael Weinreich (SPD) more specifically tied recent changes back to city-state interventions and particularly to the subsidy that was introduced back in 2005 in parts of Reiherstieg. He argued that subsidies for student housing have been implemented in neighbouring Veddel and elsewhere in the city. And yet:

> Nowhere has it influenced cultural life like it has in Wilhelmsburg. In the past, when a shop on the ground floor went under, it was always a Turkish cultural association, or yeah, mostly that, or the twentieth döner shop went in. And lately the stuff we have like Deichdiele, Kaffee Klappe, Williams Burger... it is a totally

different quality of local amenity or infrastructure that wasn't here before. And that functions only through how other people have simply come to the neighbourhood and sort of independently, without doing much, simply built their own social environment" (Michael Weinreich, interview, 2018).

Figure 37: Looking northwards on the east side of Veringstrasse, you can see some of the restaurants and cafés that have opened in Wilhelmsburg-Reiherstieg in recent years

(photo by author, 2017)

I would add several things to this analysis, including that the building of "their own social environment" is a fraught process, as I will discuss later in this chapter, and is associated with gentrification. Further, the "scene" that is developing in Reiherstieg is partly happening through the entrepreneurship and responsiveness of middle-class racialized Wilhelmsburgers, such that the idea that new arrivals are "independently" building an environment seems inaccurate. Many of the businesses on Veringstrasse, in particular, are owned and operated by long-time Wilhelmsburgers, some of whom grew up on the island with parents who had small businesses.

Some people referred to the impact as like "Sternschanze" or a "little Schanze" referring to a hip inner-city neighbourhood on the other side of the river that is packed full of cafés, restaurants, and patios. The analogy underscores the gentrification analysis that I illustrated in the previous chapter. Sternschanze is a major destination on the weekends with an extensive restaurant and bar scene, but it is also an historically working-class neighbourhood that is presently known as both a centre of left-wing radical politics and as a neighbourhood under tremendous pressure to gentrify (cf. Krohn, 2018). Wilhelmsburgers' enjoyment of the Reiherstieg cafés and restaurants that they associate with Sternschanze, where gentrification is much more advanced, echoes research findings for example from Harlem, where Freeman (2006) found that long-time residents appreciated the amenities that gentrification brought with it, even if that it was wrapped up with "white privilege," as Freeman called it.

Other residents, like Umut (interview, 2017) found this linking to Schanze over the top, telling me "yeeeah, OK, it is just one street, that's an exaggeration!" While he used the word gentrification elsewhere in our conversation to describe the general process underway in the neighbourhood, the notion that Wilhelmsburg was really on its way to becoming a gentrifying hotspot was too much. Similarly, Murat said that Wilhelmsburg has become trendy, but the notion that it "will become like Sternschanze, that is absolute nonsense, because it is a totally different structure in this neighbourhood" (Murat, interview, 2017).

Part of Wilhelmsburg's different structure is that it continues to be an "arrival city," as Canadian Doug Saunders (2010) has influentially called urban spaces of immigrant reception. Aside from the arrival of white Germans to the island, the other recent and significant development from Wilhelmsburgers' perspectives has been the arrival of "Bulgarians." Serdar described the increase of immigration from these areas:

> For the past few years, there are other migrant groups from the Balkans, new EU countries, Bulgaria and Romania, they are also wanted in this neighbourhood. The people can speak Turkish, too, they are originally Turkish ethnic groups in their homelands. First point of reception, it is ideal for the people to come to Wilhelmsburg and get started. They also have several problems: first work, an apartment, and school. They are exactly how in the '80s Turkish fellow citizens also had problems (Serdar, interview, 2017).

Bulgaria and Romania joined the EU in 2007, and since then Wilhelmsburg has seen an increase in immigration from these countries. As Serdar indicates, many are Turkish speakers who are able to take advantage of the Turkish-speaking infrastructure developed through previous decades of migrant emplacement on the island (see Chapter 4). The immigrant group is diverse however and includes many whose mother tongue is Bulgarian, and some who are Roma.

Wilhelmsburgers' discussion of this ongoing arrival was not exclusively celebratory. Gizem, who emigrated from Bulgaria almost 20 years ago, argued that freer movement since 2014, when Germany formally opened to mobility from Bulgaria, has been an improvement for Bulgarians, who previously had to deal with family separation in order to migrate for work. Yet as Serdar indicates and I will discuss in the following section, Bulgarians are facing serious barriers to accessing the basic necessities of life, including in Wilhelmsburg. These challenges mirror those faced by migrant workers in earlier periods of racial capitalism, as I touched on in Chapter 3, but take a distinctly neoliberal form that includes so-called "self-employment" and extra-legal work and housing (Lintschnig & Schultz, 2018). As Lintschnig and Schultz (2018, translation by author) argue in an article based on in-depth research with Bulgarian workers in Wilhelmsburg, "the so-called free movement of workers has many advantages... But it also favours a shadow sector in which tens of thousands, if not hundreds of thousands, of people work in Germany" in extremely precarious and often deadly conditions.

As a result, Wilhelmsburgers disagreed about whether the arrival of more Bulgarians should be considered a "change" or the persistence of the long-standing status quo in Wilhelmsburg. Ali, who works as a counsellor for new immigrants, argued that the arrival of Bulgarians is simply a continuation of the labour migration that has been part of Wilhelmsburg for more than 150 years.

> I myself don't see any real change; I mean Bulgarians can't make a change... It is just a labour migration; they haven't played any role at all in the change of Wilhelmsburg. The actual change is by the Germans... In fact, the face of Wilhelmsburg has not really changed. It's through all the new [IBA] buildings, all the concrete blocks, that the population structure has changed. Many, many more Germans have come, and from that perspective there is a change (Ali, interview, 2017).

From Ali's perspective, the structures of Wilhelmsburg has remained the same, governed in particular by systematic precarization of workers and racialized access to citizenship.

The Wilhelmsburgers I spoke to seemed likewise not to see the arrival of refugees over the past five years as a change, which is a remarkable contrast to the national discourse since 2015 and indeed the discourse in other parts of Hamburg. At one point, more than 3,000 refugees were housed in Wilhelmsburg in camps and container housing, and Wilhelmsburgers mobilized significant support for new arrivals (see for example Die Insel Hilft e.V., n.d.). Wilhelmsburg did this work rather quietly in a context where residents of some of Hamburg's wealthiest and whitest neighbourhoods were fighting tooth and nail against the location of refugee housing in their backyards, including with absurd displays in which the site of a planned refugee shelter was blocked with boats and limousines (Patterson, 2016; Spiegel Online, 2016). Other Hamburgers created city-wide networks dedicated to limiting the spatial concentration of refugees in the name of "mixing" to promote "integration" and "sustainability" (Hamburg für gute Integration, n.d.; Initiativen für erfolgreiche Integration, n.d.). The initiative was soundly rejected by prominent Wilhelmsburg activist organizations such as Zukunft Elbinsel, which argued that the fundamental human right to seek asylum should not be up for debate in any way (Zukunft Elbinsel Wilhelmsburg, 2016).

Where the arrival of refugees to Wilhelmsburg did come up in conversation was in relation to the struggle to find affordable housing. Wilhelmsburg used to be a refuge in this respect, a place where affordable apartments were always to some extent available. As Metin Hakverdi (interview, 2018) put it, Wilhelmsburg's housing market used to be practically decoupled from that of the rest of the city, with higher vacancy rates and generally lower rents. Yet housing was far and away the top concern that I encountered among Wilhelmsburg residents and their concerns were echoed by local service providers and politicians. The cost of housing (rental and ownership) has been going up and it has become much more challenging to find an apartment at all. From residents' perspectives this is tied directly to systematic discrimination against "migrant" apartment seekers. "The Bulgarians" in particular are dealing with homelessness, overcrowding, and predatory landlords. In the following section I turn to this aspect of change on the island as residents viewed it.

Figure 38: An empty lot in Korallusviertel, one of the visible signs of how certain parts of Wilhelmsburg had been forgotten while others were under redevelopment

(photo by author, 2017)

Figure 39: One of four sites of longer-term, yet still temporary housing for refugees, located near Veringkanal in Reiherstieg. The area is much smaller than at the reception camp and there is no guard at the gate, but there is still a fence marking out the cluster of containers

(photo by author, 2017)

Figure 40: The refugee reception camp on Dratelnstrasse seen in 2016. The camp was separated from other, residential parts of Wilhelmsburg by some distance, fencing, and a guard at the front gate, as is typical of the camps

(photo by author, 2016)

Figure 41: In 2018 the camp was gone, leaving an empty expanse where the container housing had been

(photo by author, 2018)

Figure 42: A few containers remained, apparently administrative offices, some with information still posted in English on their exteriors. One rain- and sun-battered sign read "There are about 800 people in the camp, please follow the house rules for your and your neighbour's safety!"

(photo by author, 2018)

Figure 43: These stacked red containers are also refugee housing, located in an industrial area that is slated for redevelopment into the Spreehafenviertel

(photo by author, 2017)

Housing in Wilhelmsburg: Expensive, scarce, and exclusive

What I heard consistently in my interviews with residents was that housing in Wilhelmsburg has become more expensive and less available, especially for racialized people. Wilhelmsburgers had noticed changes in housing costs both for renters, who are the vast majority on the island, and for those who want to buy an apartment or house. Nalan's parents, for example, bought a house in 2007 that doubled in price by 2017. "Our apartment, too, is twice as expensive now," she said. "Our neighbours are selling their apartment, for example, 110m², they are asking €300,000 and they bought it for €160,000... That is really expensive for Wilhelmsburg – but they go immediately. They're sold right away" (Nalan, interview, 2017). Mohammed (interview, 2017) has noticed the same thing, and said, "it's great, of course, for those who have it;" but for those who do not, it is a different story. Özden regrets that he did not buy an apartment years ago when prices were lower. As his family has grown, they have struggled to afford sufficient space; they used to live in neighbouring Veddel, but moved to Wilhelmsburg once family-sized apartments started to be turned into student housing in Veddel. Student subsidies were implemented in Veddel even before Wilhelmsburg, as I mentioned in the previous chapter. "On the one hand, they are also right – they all said 'we want to change the structure a bit.' I agree, but as a father I also have the right [to stay] in the same place – for families it is important, if they have kids, school et cetera ... to move to another place was complicated for me" (Özden, interview, 2017). He now lives in an apartment that his family likes, but that costs more than he can really afford, and without prospects to buy in the future or even to move to a more affordable place. "What can you do? For family and kids, you have to swallow it" (Özden, interview, 2017).

Affordability and sheer availability have become real problems for renters in Wilhelmsburg. Paula for example has an apartment she likes, but her son has been looking for a while and wants to stay on the island with his friends and family.

> At the moment, everything is super expensive in Wilhelmsburg, I find… He doesn't earn so much either, he earns maybe €1,500, and so the rent can't be too much either. And he's looking now. He has a room right now. They offered him a room with a little kitchen for €600 inclusive, I think that's an awful lot when you only earn 1,500. But it's like that for me too… half of my salary goes to rent (Paula, interview, 2017).

There is also a lot of competition for any apartments that are available. Pastor Obeng has been looking for an apartment for more than a year while staying with a relative in Kirchdorf-Süd. He told me that "There is a lot of apartments you can go and view, but they do not offer it to you. Because sometimes when you go view one apartment, you can see about 100 people going to view one apartment. And that is what has been going on: you go and look at an apartment, you do not hear from them again" (Peter, interview, 2017).

This is a familiar story in many cities, including my hometown of Toronto, but it is a very new one for Wilhelmsburg. Local politicians acknowledged the change and the strain it puts on residents. Michael Weinreich (SPD, Hamburg Parliament) described how people come to his office looking for help finding an apartment, something that was easy to do ten or fifteen years ago. "Well, that has changed, of course," he told me. "This is something that makes people feel insecure and is difficult for many people. Because people come in and say 'we've had a baby, now we have two, and we're still sitting in a two-bedroom apartment, yeah. We do not know what to do now'" (Weinreich, interview, 2018). Gülhan, whose job is specifically to support people to meet their basic needs on the island, finds that she can hardly help when it comes to housing. "Apartments are *still* a problem, always. In 2019 they're supposed to start building, and one hopes for that, there in Dratelnstrasse…They built for the IBA, but they built these apartments that – €1,500 – the residents here can't afford that…More apartments have to be built, affordable housing, subsidized housing" (Gülhan, interview, 2017).

This was a common refrain from many of the racialized Wilhelmsburgers I interviewed: more affordable housing is urgently needed, and residents looked to city politicians to take the lead. Bayram Inan (*Die Grünen*, member of the district assembly for Hamburg-Mitte), said residents often give him quite an earful on the topic.

> I hear from many citizens, if I stand at the info booth here at the market, some people complain: 'Hey, Mr. Inan, what you are all doing politically is such bullshit, it is unacceptable. Apartments [are] expensive, there are no apartments...' Or migrants complain that they aren't getting apartments. On the one hand, they are right, because some officials are narrow-minded, or they think so nationalistically, they say 'the Germans come first, and when no German wants it, then you can have it.' That can't be (Inan, interview, 2018).

What Inan hears from Wilhelmsburgers illustrates how increased cost and decreased availability intersect with discrimination against "migrants" to make it particularly hard for low-income racialized people. Inan refers to narrow-minded housing providers, but the policy of social mixing specifically asks the gatekeepers of public housing to think "nationalistically" in the name of mix, and to use racialized displaceability to exclude certain people, as I demonstrated in the previous chapter.

It therefore seems that racialized Wilhelmsburgers are feeling the direct and indirect results of social mix policy. Gizem said for example that SAGA, Hamburg's public housing provider, has the only affordable options, but "at SAGA you find 99% discrimination" (Gizem, interview, 2017). She has never heard of a Bulgarian like herself getting an apartment at SAGA; people are always told "we don't have anything." To illustrate how she knows this is discrimination, she told me of an investigation by the Hamburger Morgenpost, in which it submitted the same housing applications with a Turkish name and a German name, and found those with German names were much more likely to be invited to a viewing (see Gözübüyük & Weinhold, 2017; Schäfer, 2017). Casim likewise perceived that "Turks, foreigners, aren't getting any apartments. If they want to move here, if they look for a rental apartment, they don't get it." I asked him why he thought this was, and he replied: "I have no idea. It is – they want it mixed here, not so much from one side" (Casim, interview, 2017).

The outcomes of this kind of housing exclusion in Wilhelmsburg were often described to me as quite dire. Practically everyone I spoke to was either stuck in an apartment that in some way did not meet their needs, but from which they were unable to move for lack of other options, or had friends and family who were stuck in such a situation. Often this was a question of space – another bedroom was needed to accommodate a growing family, for example, but was not possible. Other times it was a question of condition: Gizem's place desperately needed renovation, but she was resigned to living with it, as prices for one-bedroom apartments have jumped too much for a move to be realistic. Meryem's place likewise is cold and heating it is expensive because of old insulation, and she waits and hopes that the landlord will get around to doing something about it.

The worst conditions are faced by racialized people who have the lowest incomes. Mostafa told me for example that he knows of people who sleep in garden sheds, even in the wintertime.

> Sometimes people are tense because they live in apartments with ten people. With ten people in a two- or three-room apartment! It creates complications in the family. The income level is very low, because there are a lot of unskilled people, to an extent, who live in this neighbourhood. They don't earn much, so they just have to share an apartment with other families ... they are for example people from Bulgaria, EU citizens... we have a workers' market in the mornings (Mostafa, interview, 2017).

In fact, there are multiple early morning workers' markets in Wilhelmsburg, where nominally self-employed immigrant workers sign up for a day's labour. Wilhelmsburgers are aware that Bulgarians are often paid well under minimum wage to do dirty and dangerous work like cleaning and construction; the scarcity of housing is exploited by landlords who squeeze people into cellars, attics, and single rooms and then charge €500-600 per person for the privilege (see also Lintschnig & Schultz, 2018). Lately there have been regular reports of illegal housing raided and shut down, often in spaces that are not zoned for residential use (cf. Eicke-Diekmann & Popien, 2018). There are also Bulgarian and Eastern European Wilhelmsburgers who live in cars or largely out of doors, in tents and makeshift structures in local parks. In September 2017, a man from Bulgaria was found dead in a Wilhelmsburg park, having succumbed to illness without health insurance and without a home after 11 years of working in Hamburg (Lintschnig & Schultz, 2018).

A central aspect of peoples' vulnerability in this situation is that housing is a key to life and life chances in Germany. As Gizem pointed out to me, one needs an address to register with the city, and registration is necessary to work legally, to send one's kids to school, and to get a bank account. If you do not have a valid local address you thus have multiple problems on top of homelessness or underhousing. At the same time, to get an apartment, you need to show proof of income, which you likely cannot get if you are working informally. Further, health insurance is typically accessed through one's employer, and requires a valid contract. People get caught in a "vicious circle," as Nalan put it, eking out a living outside of legal legitimacy. She marvelled that Eastern Europeans stay in Germany under these conditions. "They say it is better here than there anyway, and so imagine what the situation must be like there. That's really extreme" (Nalan, interview, 2017). She also wondered why the various levels of government do not intervene in this vicious circle to change policy to prevent people from such distress. She concluded that they must know and not care, a perspective that was shared by Ali, who viewed the vulnerability of some Wilhelmsburgers as fundamentally structural in nature, tracing back to legislation that creates and maintains inequality and differential citizenship at the national and international levels, as I also discussed in Chapters 3 and 4. Viewed through the lens of racial capitalism, in which the production of value requires differentially valued bodies, it is also not an accident or mistake that some people, in this case nominally European "others," struggle in a state of semi-legality and experience exploitation in the gap. This state of semi-legality, along with racialization as non-German, facilitates the extraction of value from the labour of immigrant workers.

Figure 44: One often sees signs that people are living outdoors in Wilhelmsburg. Here it is bedding under a tree by the Ernst-August Kanal in Reiherstieg

(photo by author, 2017)

Despite their critical analysis of the current housing situation in Wilhelmsburg, Wilhelmsburgers were often quite enthusiastic about the very practice of social mixing that they implicated in the discrimination and vulnerability that they problematize. The notion of a good or healthy neighbourhood mix had a lot of support among the racialized Wilhelmsburgers I interviewed and underpinned their approval of the recent arrival of more white Germans to the island. In the following section I explore this tension, arguing that Wilhelmsburgers have "mixed feelings" about mix, again drawing on Haritaworn's (2012) phrasing in their exploration of the perspectives of people with part-Thai heritage in Britain and Germany. Wilhelmsburgers reject the discrimination to which social mix is linked, yet hold out hope that it undermines the processes that have stigmatized and isolated Wilhelmsburg to date. I complicate this picture further in section five, in which I explore residents' analysis of conflict and tension in the context of mix and conviviality.

The promise and illusion of social mix

As some quotes from residents hinted in the previous section, many of the residents I interviewed argued that the "mix" in the Wilhelmsburg population had been changing steadily in recent years. In general, they felt that this development was a positive one, something that "is going in the right direction," as Serdar (interview, 2017) put it. Umut strongly agreed, for example, saying:

> Overall our neighbourhood is really positive, very mixed at the moment, and I hope that it stays mixed, and the development has, I believe, started – there are even more building projects that will probably bring more families to Wilhelmsburg, but also student residences that will bring younger people. I hope that this mix between socially disadvantaged and socially respected people will remain, so that the mix will be better secured (Umut, interview, 2017).

Wilhelmsburgers' enthusiasm for mixing seemed to be tied to the prevailing sense that mix had gotten lost over previous decades on the island. Residents who had lived in Wilhelmsburg the longest described a past in which they, as "foreigners," had been in the minority in their apartment buildings and schools and argued that those were better times from a mixing perspective.

Some Wilhelmsburgers placed the responsibility for the un-mixing of Wilhelmsburg, as it were, squarely on the city-state and its planning decisions. I quoted Arzu in Chapter 3 on what she sees as the deliberate concentration of an "underclass" on the island and Mohammed argued as well that the city had made an enormous mistake by concentrating social housing in Wilhelmsburg. He also related it to other neighbourhoods with concentrations of social housing. "That meant in principle that they brought a circumscribed social class here, and not the upper strata but the lower… That was really really, really, really bad housing policy, it is basically what they should never do, but you have that in Billstedt, Steilshoop, Mümmelmannsberg, Wilhelmsburg" (Mohammed, interview, 2017). Paula described the loss of a past mix as a process that took place over time.

> I think this mixing ratio should have been better. I find – I like multicultural, so maybe like it was at the beginning of the '80s or '70s. I found it better when it was still, maybe 50/50, that was a good mix. It is just more Germans who aren't

> doing well, they stayed here, the ones who couldn't manage to build a house or to move away, they stayed here, and the foreigners stayed here. That is just maybe not so positive, it didn't develop so well (Paula, interview, 2017).

On the question of *why* that was not a "positive" development, Wilhelmsburgers' perspectives reflected some of the common sense of mix in planning that I identified in the previous chapter. Mohammed argued, for example, that homogeneity is a problem in and of itself that leads to the intensification of issues associated with a particular social group or class. To illustrate this, he jokingly asked me to imagine the kinds of conflicts and drug use that would plague an all-academic neighbourhood. He particularly felt that class mix helps to motivate youth from less educated families, as it offers potential that they will find other role models and differently imagine what they can do or be.

Other Wilhelmsburgers hoped that increased mix in the neighbourhood and in particular buildings would mean more contact with white Germans. As Casim (interview, 2017) put it, "I find mixed is better, because, yeah, if one lives in Germany, one can also be friends with Germans, talk to them, too." Indeed, some of the racialized residents I spoke to seemed to long for that but found it a struggle to achieve. Bayram Inan and Meryem both told me stories of passing their German neighbours in the stairwell for years before they got any acknowledgment. Meryem described herself as shy, but Inan in contrast related "they never said 'hello,' never said 'good morning,' and I would purposely say 'Good morning, how are you?' Even on Christmas I would congratulate them, visit, bring gifts. At some point they gave in" (Inan, interview, 2018). He and others felt based on their own past experiences that proximity would eventually lead to contact and neighbourly relationships even if it took a good deal of time and one-sided effort. Inan thus said that "Turkish people tell me, 'Yes, Mr. Inan, they don't want us.' No, it doesn't matter if they want us or not. We have to always make the opening: 'Hello.' I say hello ten times, it doesn't matter if he says nothing, at some point he will give up" (Inan, interview, 2018).

For some racialized Wilhelmsburgers, the promise of mix is thus the potential to achieve *conviviality* with white Germans by pushing them into the neighbourly relationships that they have avoided or fled in the past. Conviviality refers to processes of cohabitation and interaction that make multiculture an ordinary part of life (Gilroy, 2004), as I discussed in Chapter 4. Wilhelmsburgers' interest in mixing from this perspective underscores Gilroy's (2004) argument that conviviality does not mean a triumph of tolerance necessarily, or an absence of racism. Contentious power relations remain part of the picture when multiculture is understood as an active practice that white Germans need to be encouraged to undertake. While mixing was framed by some as correcting the city's past urban-planning mistakes, others framed it as potentially correcting the mistakes of white Germans themselves.

> It is a bit of a strange feeling; we have remained between cultures. I'm Turkish, I live in Germany, but I can't be among Germans. I can, but Germans don't like to be among us. At least not that I have seen ... that is why the IBA saw that the structure has to change, to bring the Germans in. Students are ... one can't say weapon, but the student is an important way, because students need apartments and affordable apartments (Özden, interview, 2017).

For Özden, the city's most visible recent interventions look like strategies to encourage and persuade Germans to come (back) to Wilhelmsburg. Instead of framing this as an "immigrant" problem or failing, however, as the dominant discourse about Wilhelmsburg and problem neighbourhoods does (see Chapter 3), Özden frames this as a German problem: they have taken too long to accept having neighbours from Turkey, and need a push. Paula similarly argues that mix needs to be cultivated and maintained, not because Wilhelmsburg is homogenous, but because Germans cannot handle when the population of "foreigners" gets too large.

> This xenophobia has increased. Wilhelmsburg is still, I think, not quite as bad as in some other neighbourhoods, because here there are a lot of foreigners: the majority here are foreigners. But a lot of Germans are just annoyed by that, to the point of being xenophobic. Sometimes they tell me, they complain about particular nationalities or whatever, and I say 'yes, but I'm a foreigner too,' and they say 'not you.' I feel attacked sometimes, and then I think again OK, they don't mean it personally. But I don't think it is good that they mean it at all [*laughs*]... I think, yeah, it isn't justified to always say things about all people, or all foreigners or all Turks, all Italians: no one is 'all' (Paula, interview, 2017).

Paula argued that her experiences are different from many other Wilhelmsburgers, as she is able to pass for "German" in certain situations, because her name, appearance and religion do not immediately mark her as "foreign." Paradoxically, she is exposed to German racism because she passes, despite the fact that she developed strategies to blend in as a result of experiences of racism in childhood. Paula's perspective is that a certain mix needs to be restored in Wilhelmsburg in order to keep white Germans happy and not trigger xenophobia. Xenophobia is a term that is often used instead of "racism" in the German context, though in practice many of the people targeted are not "foreigners" at all, as previous chapters have argued. A tension in this analysis is that she argues at the same time that racism is worse in neighbourhoods where there are fewer "foreigners," a perspective that is shared by other racialized residents, as Chapter 4 demonstrated.

Like the dominant discourse on social mixing in which, as argued in Chapter 5, the precise terms and goals of mixing are often unclear, Wilhelmsburgers' perspectives often brought together and sometimes jumbled the terms of mixing. Racialized residents talked about mixing classes and also about mixing "German" and "non-German," sometimes equating the two and often reifying categories and assumptions that they otherwise rejected. When it was possible, I asked interviewees what they thought about Zeynep Adanalı's (2013) findings that public housing was being "mixed" in Wilhelmsburg on the basis of income *and* "migrant-sounding names." This garnered a variety of responses ranging from cautious support to skepticism and concern with the impacts. For example, Mohammed's response was:

> If [mix] is the basis for it, I think it is totally OK. I do, actually. Because I think, if someone moves to Wilhelmsburg, then it should be a group that mixes the whole more. Although, for me it doesn't mean foreigner and German, rather it means, I would look what does the person do, and I'd try to achieve a social mix. If I looked more at lawyers and whatever with foreign names, I'd take them too... Regardless of skin colour, blond, or not, it really doesn't matter, it is about a social mix (Mohammed, interview, 2017).

Mohammed thus valued a mix that is based on class and education, rather than what amounts to race, which I noted in the previous chapter local politicians generally do as well. I found this analysis to be shared by many Wilhelmsburgers, but also tied to their own social location and economic position. One of the tensions in this analysis is that by the class and education criteria that Mohammed advocated, around half of the Wilhelmsburgers I interviewed would be considered eligible for housing in Wilhelmsburg today, and half would not. This did not represent a threat to the residents who owned their homes, or who were confident in their comfortable financial position. However, for those with lower incomes, insecure work, or health issues for example, the racialized displaceability embedded in social mix meant they were bound to stay where they were, as I noted above.

Several Wilhelmsburgers recognized risks embedded in the mixing process and were concerned about how it would continue, even as they were enthusiastic about the arrival of more white Germans in the neighbourhood. Nalan included among her hopes for the future of the neighbourhood "that it doesn't change *so* much now, that the population isn't replaced so that in 10 years it is completely different. [Wilhelmsburg] should also be kept the same, just a bit improved in terms of infrastructure. Not much more should change: why should it? It is nice here" (Nalan, interview, 2017). Umut agreed and said that he found the mix to be ideal at the moment.

> It is very close to a balance, very even, but I fear that it will just go farther in the direction of the socially stronger classes and the socially weaker will be pushed further towards the margins. First through the rents, and yeah, et cetera, also societal belonging... I want there to be a balance, that the new and old Wilhelmsburgers – that is what makes Wilhelmsburg, actually, that there are people

> there who live next to each other and don't understand each other, or not just linguistically but also culturally don't understand, but live peacefully together and alongside each other (Umut, interview, 2017).

Nalan, Umut, and others perceived a threat embedded within the development that they otherwise find quite positive. The evidence on social mix that I discussed in Chapter 1 and Chapter 5 demonstrates that the threat of displacement, and of domination by more powerful new arrivals is indeed quite real. The threat manifests in how the cost and condition of housing has deteriorated in Wilhelmsburg in recent years, creating increased hardship particularly for low-income and racialized people and most intensely for Bulgarians and others in precarious work situations. It is against this backdrop that Gülhan, a social-service provider for newcomers, argued that the notion of "mix" is just an illusion.

> We hear about it, but practically I believe it means nothing. Poor and rich mix, and migrant and local mix. I think it is just an illusion, it is a distraction for us all. I think in the meantime in this neighbourhood there are already people who live with each other – just find an apartment and move in and then there's a mix. Everyone should live where they want to, of course, shouldn't they? But I have had phone calls where [housing providers] said to me 'we don't want any EU citizens.' Where I made the call because they didn't speak German so well, and 'what nationality?' 'EU citizens.' 'We don't want them.' There's that too, it is real discrimination (Gülhan, interview, 2017).

Where others see a promising and legitimate planning tool, Gülhan saw a discriminatory illusion, based on her experiences on the receiving end of mixing strategy. "EU citizen" is a shorthand here for non-German and, given recent immigration patterns in the neighbourhood, specifically for "Bulgarians." She puts mix and discrimination into the same frame, instead of separating them as some other residents did. Wilhelmsburg and its various neighbourhoods *are* mixed from her perspective, and she thus highlights that the discourse really refers to the privileging of certain kinds of people in housing, and not to a (already long-standing) mix of different racialized people or different immigrant groups. I have tied this in previous chapters to racialized displaceability.

Wilhelmsburgers thus have mixed feelings about mix. On the one hand, the basic logic is largely supported, if with some caution. Yet the *effects* of the kind of privileging that Gülhan describes and that Adanalı (2013) has also documented, were also clearly problematized by residents. Problematizing and praise for "mix" did not break down along any clear group lines: often the very same people supported the notion of mix and lamented changing conditions in the neighbourhood within the course of one interview. Wilhelmsburgers described effects of housing exclusion that they felt to be quite serious and damaging to individuals, families, and the community, and aspects of what they described have been identified elsewhere as features of gentrification, including immobilization (getting stuck in unsuitable housing) (Slater, 2006), and inaccessibility of housing that was previously accessible to them (Hohenstatt & Rinn, 2013; Rinn, 2018).

In the following section I further demonstrate that Wilhelmsburgers' own accounts and my participant sensing of recent tensions and conflicts in the neighbourhood underscore that the kind of population mix that residents value does not necessarily equate to the kind of close living together that they wish for. There is tension between long-time residents and newcomer Eastern Europeans, who are constructed as different and deviating from local norms through discourses of fear and order in public space. There are also signs of distance and conflict between Old Wilhelmsburgers and New Wilhelmsburgers, who have distinct tastes and desires for services, as well as different positions in relation to power and resources. In the following section I explore these tensions, drawing on interviews, participant sensing, and media archives. The figures of "the Bulgarian" and the "German" again feature centrally in residents' narratives. Taken together, these two vectors of the current contestation of space in Wilhelmsburg illustrate the challenging side of "mixing:" diversity on the island, in neighbourhoods, buildings, and public spaces, is not necessarily associated with the kinds of close and friendly relationships that residents hope for, as racialization and access to resources position people quite differently.

Living with or alongside one another?

As I have emphasized throughout this book, many of the Wilhelmsburgers I met through the course of my study highlighted everyday challenges in the neighbourhood that they thought were generally overlooked, and that they wanted to ensure I did not overlook in my research. A key point that many interviewees impressed upon me is that in Wilhelmsburg a great diversity of people live overwhelmingly peacefully alongside one another. This was framed as an important, yet mundane accomplishment, a basic feature of how Wilhelmsburg is distinct from other neighbourhoods, as I explored in detail in Chapter 4 as an example of conviviality. Several Wilhelmsburgers drew a distinction, however, between living *alongside* one another (*nebeneinander*) and living *with* one another (*miteinander*). The former is what they felt that Wilhelmsburgers already do well, and the latter is an aspiration that they want to explore and pursue with their neighbours.

Wilhelmsburgers were keen that I analyze issues of power and discrimination in the neighbourhood and beyond as a complex picture in which everyone is implicated. Several of the racialized residents I interviewed took pains to explain that racism and prejudice is not only an issue of "German" versus "non-German," but also of prejudice and intolerance within communities who are themselves all "migrants," as residents put it. Murat (interview, 2017) said, "what I would like to underline is that when we talk about racism in Germany, it is really important that we position ourselves against all racisms. Because there is a lot of racism among migrants." He mentioned in particular the presence of Turkish racism, fascism, and nationalism connected to political movements in that country. This is certainly a concern for other Hamburgers as well, as I will touch on below. I was in Wilhelmsburg when Erdoğan won a second presidential term in Turkey, and the celebration on the streets around my home was large and loud. Sitting in my apartment and listening to cars driving up and down and honking was unsettling given Erdoğan's anti-Kurdish policies and jailing of academics and journalists. For example, Deniz Yücel, a German-Turkish journalist, was imprisoned in Turkey during my study, one of 250 journalists imprisoned there since 2016. He was held and tortured for a year, in what the Turkish Constitutional Court has since ruled an illegal arrest and detention (Deutsche Welle, 2019; Kaschel, 2019). While Murat's focus was not on Erdoğan supporters per se, his feeling was that not enough is being done to address racist and authoritarian attitudes in all sections of the population.

Foremost in residents' minds, however, were some Wilhelmsburgers' problematic attitudes towards Bulgarians and other Eastern Europeans. Gülhan (interview, 2017) told me for example that "migrants, guest workers, Turks, they also [say]: 'These Bulgarians, they don't behave themselves here'." This is a refrain I also heard from a couple of the people I interviewed. Indeed, the one Bulgarian who participated in my research, Gizem, did so in part to counteract what she sees as the dominant images of people from her country in Wilhelmsburg, which focus on alcohol use, begging, under-employment and under-housing. Based on her experiences as a counselor, Gülhan pushes back against prejudices and stereotypes that are held by other Turkish-speaking Wilhelmsburgers. She tells people in her social circle, "I see this quite differently, [I see] the conditions they're facing, the kinds of situations they have to live with" (Gülhan, interview, 2017). She encourages people to notice the similarities rather than differences between themselves and Bulgarians, especially the shared immigrant journey, but she wonders whether people just always target the next group that is "below" them.

The spatial dimensions to Wilhelmsburgers' attitudes suggest that it is not just about prejudice but about the use of public space. Nalan noticed, for example, that Bulgarians are appropriating public parks in a particular way – cooking and eating outside, spending time outside, and sometimes sleeping outside, as I have noted and depicted above. Nalan connected this to what she considered a real lack of *private* space for people who are living in overcrowded and precarious situations, in the context of a broader housing crisis. But Yücel simply criticized how Bulgarians use the streets and sidewalks: "it does not look good, how they present themselves. I'm a foreigner myself, I don't have anything against them, but ... they block the streets, they beg, that is the biggest problem. They beg a lot ... it's not a nice picture" (Yücel, interview, 2017). Some interviewees thus constructed the figure of the Bulgarian as different and problematic because of how they use public space, and while concern about the well-being of people living precariously and in poverty was sometimes part of the narrative, sometimes a concern with acceptable behaviour in public space arose first.

Fear of crime played a role in this construction, particularly of Bulgarian men. Meryem and Zehra spoke for example about a bar that had opened in the Bahnhofsviertel, where the patrons often spilled into the street and blocked the sidewalk. They said it made them feel uncomfortable, like they could not walk on that side of the street, and Zehra worried about her daughter who came home that way from her work at a hotel downtown often late at night. Casim, who lives in Reiherstieg, also mentioned his children. "At night, when you go out, everywhere there are Bulgarians and Romanians sleeping in cars, and naturally that creates fear. For example, my kids, when they come home ... they're afraid. Naturally. They are sleeping there; one doesn't know if they are drunk or what... I haven't heard about anything happening, but you never know" (Casim, interview, 2017).

A certain sense of the unknown and uncomfortable is a feature of urban life (England & Simon, 2010), but streets and parks are stereotypically fearful urban spaces particularly because they are inhabited by the homeless (Pain, 2000; Shirlow & Pain, 2003). Homeless bodies tend to be treated as uncontrollable bodies that need to be removed in order to maintain order (Kawash, 1998). In her work on homelessness and the production of the public in the U.S., Samira Kawash (1998) argues that homeless bodies are excluded from civilized space in order to maintain an illusion of the public sphere as unified and free of conflict. From this perspective, Bulgarians or Romanians begging on the street, or sleeping in cars, destroy the illusion, making poverty, under-housing, and exploitation visible. This notably hinges in part on mobility, which as I explained in the Introduction is a central theme in racialization in Germany (El-Tayeb, 2016; see also van Baar, 2017); because under-housed, Eastern Europeans are constructed as too mobile, indeed literally living in forms of transportation.

Discourses of fear and discomfort about "the Bulgarians," whether as a generalized and imagined figure or as actual people from Eastern Europe, contribute to the definition of them as distinct (and different) people in Wilhelmsburg. Fear and fear discourses "are power-laden as they work to define and maintain the shifting boundaries between deviance and belonging" (England & Simon, 2010, p. 204). The dominant discourses about Bulgarians in Wilhelmsburg hinge on deviance. By citing fear in relation to "Bulgarians," some of the Wilhelmsburgers I spoke to articulated them as an "other" that challenges a pre-existing social order (Shirlow & Pain, 2003). Yet Wilhelmsburgers also often did this in a self-conscious way, by for example repeating stereotypes of "Bulgarians" as alcoholics and criminals, yet questioning the generalization in the same breath. As Mithu Sanyal (2019a) noted in relation to racism in the quote with which I opened Chapter 4, Othering is not the ultimate point here, rather it is rights and resources that are at issue, including who can and should use what space and how. Perhaps, as with other exercises in Othering, this is also a practice of self-definition for racialized Wilhelmsburgers who do not sleep in cars, or outside, or earn their money by begging. Against this Other, it is possible to claim respectability to which racialized people have very limited access in Germany, as I discussed in Chapter 1.

Interviewees' perspectives on "Bulgarians" also bring the nature of "differential racialization" to the fore. As Laura Pulido explores in her work on the history of Black, Chicano/a, and Asian radical activism in Los Angeles, different groups are racialized in distinct ways, and "a particular set of racial meanings are attached to different racial/ethnic groups that not only affect their class position and racial standing but are also a function of it" (Pulido, 2006, p. 24). The material conditions facing many Bulgarians in Wilhelmsburg, especially the intensity of exploitation that is tied to their particular location at a specific (European) political and economic juncture, affects the racial discourse about "Bulgarians." The racial discourse, in turn, likely feeds into notions of exploitability by employers and landlords.

Residents' mentions of fear in relation to Bulgarians must be situated as differential, reinforcing of existing social relations, and constantly constructed in ways that are gendered, aged, classed, and racialized (Pain, 2000, 2001; Shirlow & Pain, 2003). It should not surprise that interviewees suggested that the Bahnhofsviertel has certain streets that do not feel safe to them, considering that women have well-founded fears of men, and people report more fear of crime in relation to spaces that are perceived as neglected or unattractive (Pain, 2000). Yet at the same time, people who are objects of fear are often actually those most vulnerable to violence themselves [there were 1,713 attacks on refugees and refugee shelters in Germany in 2017, for example (Pro Asyl, 2017)]; this is particularly true for people who are homeless (Pain, 2000; Shirlow & Pain, 2003). As I have also touched on in previous chapters, the supposed danger of particular men to "our" women and to LGBTQ people – whether it is Black men, Muslim men, Syrian men, Bulgarian men, depending on the context – is often at the centre of racialization and racism and thus the definition of certain people as threats (Pain, 2001; see also Hall, 1997). Fear discourses tend to be amplified by the media, and are mobilized to pursue political goals (Stjernborg et al., 2015), which Wilhelmsburgers themselves analyzed in relation to the island's history, as I demonstrated in Chapter 3.

Wilhelmsburgers Old and New

The residents I interviewed also reported some conflicts and tensions between "Old Wilhelmsburgers" and "New Wilhelmsburgers." Some people in the neighbourhood do not appreciate the distinction between Old and New, finding it divisive or exclusive of newcomers to the island. I found, however, that New/Old was a way in which the long-time Wilhelmsburgers I spoke to talked about race and class without naming it directly. In the context of my study in particular, I understand it as a way in which racialized Wilhelmsburgers for whom the island is *Heimat*, as I discussed in Chapter 4, continue to assert their long-time presence in relation to the changing population. Racialization remains important because of how racism has shaped the island and residents' lives to date. Crucially, I never heard Bulgarians or Eastern Europeans included within the notion of the New Wilhelmsburger, which underscores my analysis that New/Old is about class and racialization rather than date of arrival. It is the figure of the "German" that aligns with Newness in this sense, as is illustrated in the nature of the spatial and social divides that my interviewees highlighted.

Residents of Wilhelmsburg argued, for example, that there are aspects of spatial and social divides between New and Old Wilhelmsburgers that are observable in daily life.

> The best example is I think if you're in a shop, in a café, that has a hipster-ish design – maybe I won't mention any names [*laughs*] – you don't see any, I would say, people with roots in Turkey or people with other roots who have lived here for generations. They probably don't feel – they think the shop is fancy, like everyone, find it somehow different, like everyone, but they wouldn't venture to go in because they'd just be in the minority inside and would feel uncomfortable, because they also wouldn't have anything to talk about with the other people there (Umut, interview, 2017).

This cuts both ways, in Umut's estimation. The weekly market is known for its large selection of inexpensive clothes, a specialization that has grown over time. It is more Old Wilhelmsburgers who tend to shop there, he argues.

> But if you say there is a food truck festival, at the same site [Stübenplatz], it's more the New Wilhelmsburgers there, in the same place. That was for example last year, and long-time residents were not there. Not many, they checked it out, obviously they are in the area every day, but they didn't go and say 'OK, hey, I'd like a burger,' there, or sweet potato fries [*laughs*] (Umut, interview, 2017).

This has to do with the subtle and sometimes not-at-all subtle ways in which people feel that something is for them or not for them. Prices are certainly a large factor, but as Umut notes it also has to do with feeling like something fits you and appeals to you in tangible and intangible ways.

The question of who feels comfortable and welcome where is something that Umut is particularly interested in. His family, who have owned and operated a convenience store in Wilhelmsburg for decades, recently opened another business, a beverage shop with an explicit intention to appeal both to New and Old Wilhelmsburgers. The strategy, as he framed it, is to carry products that both markets might want: international, craft, and upmarket beers, as well as soft drinks and the least-expensive brands of beer. Inherent in this strategy is a recognition of different budgets and of different needs and tastes; several Wilhelmsburgers argued that any business that opens on the island now needs to recognize this to be successful. Umut jokingly called this a mission of bringing the changing neighbourhood together over drinks. Yet I noticed an intangible side to the shop's success so far, as I became a regular customer while I was living around the corner: the people who work there are all multilingual, many have been present in the neighbourhood for many years. It is not only the products that appeal to a range of people in the neighbourhood, but the people who are working there and are engaged in everyday neighbourly relations with a range of different people. This reminds me again of the importance of normalized representation of racialized people in the island's public life, as I discussed in Chapter 4.

Yet the residents I interviewed reported instances of displacement of Old Wilhelmsburgers from the commercial landscape of the changing neighbourhood. Pastor Obeng's congregation is currently struggling, for example, because they do not have a permanent space where they can meet and store the instruments that are an important part of their church services. They used to have their own building, but were moved out so that it could be used for offices. "We rented the place from a different company, and that company sold it to this company. When this company came in, it was a very large place but the amount we were paying seemed to them very little. The place could not be used for offices, because it was a large auditorium. So maybe they decided to divide it into small offices and make money out of that" (Peter, interview, 2017). I also heard the story of a family that has had a small business in Wilhelmsburg for over 20 years, and is now being squeezed out of their rental space. In 20 years, their rent has gone up exponentially, and recently the landlord declined to extend their lease, arguing that the shop's *Konzept*, which translates roughly as its concept, idea or vision, is no longer a good fit.

> A 'concept' didn't use to be necessary in Wilhelmsburg [*laughs*]. As a landlord you were just happy if a shop wasn't standing empty. You just went in. Now they want to have a concept, and on the grounds that the concept didn't fit the shop or the neighbourhood now [their lease wasn't extended]. So, I ask myself, OK: two years ago it fit, 20 years ago it fit, why doesn't it fit now? … This is a story that

Figure 45: The contestation of public space was often visible in Wilhelmsburg in garbage-related activism. In the above photo, someone has tied a homemade garbage bin to the trees in the park in front of my building. They read "Respect nature, respect yourself" in English, suggesting that they are aimed at a non-German-speaking public. The following year, Hamburg city services has installed their official, red garbage cans in the park

(photo by author, 2018)

> really shook me and that makes me really sad... They pay their rent on time, they pay every rent increase over the many years, it is outrageous to do such a thing, it is morally outrageous ... it is always this greed, always this more, more, more (Umut, interview, 2017).

Umut's analysis is that this story is ultimately about a building owner's desire to maximize profit, regardless of the cost to a family's livelihood and future. He is disturbed by the willingness to push a family out specifically when it is possible that they will have nowhere else to reopen their business in the changing neighbourhood. Yet "a fitting concept" must also be understood in direct relation to the arrival of white Germans specifically, since as Umut asks himself: why else would a small, migrant-run family business suddenly not fit the neighbourhood? Migrant-run small businesses have long been part of Wilhelmsburg's charm, according to locals like Leyla Yenirce, a journalist who wrote a year-long weekly column from Wilhelmsburg. Some of this charm includes things like staying open late and drinking

Figure 46: Garbage activism is also visible on the streets, here in front of the drugstore on Vogelhüttendeich in Reiherstieg

(photo by author, 2018)

coffee over computer repairs (Yenirce, 2017b), and other practices that challenge dominant, white, German business norms.

Just because Wilhelmsburg's demographics are changing does not mean that everyone is mixing or getting along, an observation that is also reflected in the critical literature on social mix, as I noted in the previous chapter. Arzu told me for example about the separation of white, German kids from others in her daughter's daycare. Her daughter was moving from one age group to the next and Arzu was told to put her in a particular group.

> Then at some point I saw this group, and of 16 kids, 13 were Turkish. And then I asked: 'what is this supposed to be?' [*laughs*]. And there was just one German kid there, an African kid, a Spanish one; so only migrants. And I asked what that was supposed to be, because there are groups where there are only German kids. I'm not kidding. Then I raised such an outcry, where I said, 'my kid is not joining this group'– not because I have a problem with it, if all the groups were like that, it would be no problem for me... But the other group, that was mainly German, stayed mainly German. Then the preschool director said to me, 'the parents demand that one kid be with another and that's how it develops.' They have to stop that. Listen: you can't always get what you want! (Arzu, interview, 2017)

Arzu told this story to illustrate how early in life discrimination begins for racialized people. It also reminds me of urban sociologist Jan Wehrheim's (2018) observation that middle-class people tend to shape space in their own image. That image is specifically white, they make daycare groups reflect their racially segregated social groups, which is indeed not uncommon in German schools (Karakayali & Nieden, 2013). The dynamic also seems to reflect the persistent devaluation of Wilhelmsburg and the people who have lived there for a long time. Arzu argued elsewhere in our conversation that the influx of resources to Wilhelmsburg seems intended "just to make it palatable to those who have newly arrived, to New Wilhelmsburgers, so that their expectations are met, and they don't move away again" (Arzu, interview, 2017). While she was speaking there about the "support" that the island is finally receiving after decades, something similar perhaps also applies to daycares and other institutions that want to keep white, middle-class Germans happy.

Tension or conflict between the interests and perspectives of New and Old Wilhelmsburgers has also been made evident by some recent leftist activism. On May 1, 2017, traditionally a day of labour and anti-capitalist demonstration in Germany, a group marched through Wilhelmsburg-Reiherstieg and broke the windows of a vegetable shop, convenience store, kumpir restaurant, as well as the Deutsche Bank, all on Veringstrasse. The next day a group claimed responsibility on the leftist website *Indymedia Linksunten* (a website that has since been banned by the German government as part of post-G20 anti-leftist measures) calling it a spontaneous action in solidarity with the Kurdish resistance. They explained their targets as shops owned by known supporters of Erdoğan and the AKP in Turkey, as well as a restaurant (the kumpir place) that had been "actively advancing gentrification of the neighbourhood" (Anonym, on Linksunten HH, 2017, translation by author).

Figure 47: The bank's windows seem to have suffered the most damage. Seen here alongside posters that read "For the G20 summit in Hamburg: break the chains, paralyze the port"

(photo by author, 2017)

The action and its rationale garnered some support online, but also critique and requests to the activists to reflect upon the impact of their actions. "This is meant to be a critique in solidarity: What remains for many people in the neighbourhood after this action is broken Turkish convenience stores and vegetable shops. People haven't gathered that it was about the AKP or MHP [a far-right party in Turkey], rather many think of the attacks as a right-wing attack in times of the NSU, AfD, etc" (*Bitte Nachdenken*, May 3, 2017, Linksunten HH, translation by author). The critique was accompanied by suggestions that in the future, actions should make the message much clearer, not only to targets but to local residents; others argued that the small businesses of local racialized people are entirely unacceptable targets if demonstrations are to be anti-racist as well as anti-fascist and anti-capitalist. The debate became more heated as it became clear that those who had undertaken the action were not taking friendly feedback well. One exasperated forum participant wrote that

> It is about the people who live here! For the last few years Germans are moving here in large numbers to an immigrant neighbourhood, and think that just

Figure 48: The kumpir shop as seen from Williamsburger across the street. In the window on the right you can see damage in the burger shop's window, I believe also from May 1st; they have repaired but left it visible, crowned with a smiley face (also see Figure 36)

(photo by author, 2017)

> because they're punks and leftists they can do anything, because their political perspective legitimates any action and their gentrification as well. There is no visible integration on your parts! No connection to existing structures, or coming together, or approaching people that have lived here for a long time ... Wilhelmsburg is not the Schanze you idiots, and if you claim to want to be in solidarity with Kurdistan, organize yourselves with the Kurds in the neighbourhood a little better and let them smash the shops. You could have also smashed the fucking AfD that's established in Wilhelmsburg... (Ihr in eurer Blase, May 4, 2017, Linksunten HH, translation by author).

In her *taz* column, Kurdish journalist Leyla Yenirce, herself a recent transplant to the island, wrote a few weeks later about the consequences of the action from her perspective.

> The fact that the appropriation of the Kurdish freedom struggle through throwing stones at Turkish businesses only furthers the criminalization of Kurds seems to be as irrelevant to the activists as the fact that the kumpir shop is run by a family that has lived in Wilhelmsburg for generations... The owner's roots are in Afghanistan. It probably would not have occurred to her when she opened her shop that she would be a political symbol of the downfall of Wilhelmsburg. Not least because the neighbourhood was already adorned with a café with vegan

> cake and a leftist shop with Fair-Fashion and organic production (Yenirce, 2017a, translation by author).

It is the shop owner herself who paid the price for the demonstration, through the cost of cleaning anti-gentrification graffiti off her storefront. In fact, in the time I spent in Wilhelmsburg in 2017 and 2018, only the kumpir shop and the bank had the damage repaired; the windows of the convenience store, and those of the vegetable shop, remained broken and covered over by tape. The failure to understand and take seriously the impact on locals, and the focus in particular on "Turkish" fascism rather than the AfD, illustrates the underlying racism in some leftist activism. Other leftists are putting more intention into anti-racist work in and about Wilhelmsburg, but this was a highly visible problematic example.

The incident and surrounding debate illustrate some of the contestation of Wilhelmsburg in the context of change. Umut referred to "New Wilhelmsburgers who think they are Old Wilhelmsburgers," to characterize how some relative newcomers to the island assert the supposed interests of island residents, though they do not really know Wilhelmsburg in depth. As the forum post above put it, white German arrivals have not yet integrated themselves into Wilhelmsburg; this would require actually connecting with long-time residents. Instead, from some Old Wilhelmsburgers' perspectives, it looks like they see their vision of the island as the only real one.

The protest and debate also draw to attention how long-time Wilhelmsburg residents are using the opportunities that the "mixing" of the neighbourhood presents for them. Many of the newer businesses in Reiherstieg – cafés, restaurants, and shops – are in fact owned by long-time Wilhelmsburgers, often racialized Wilhelmsburgers whose parents have had small businesses on the island for decades. Participants in the community event I held at the close of my data collection (see Chapter 2), told me that this reflected the hopes of residents who participated in the Future Conference back in the early 2000s, namely that city-state interventions in Wilhelmsburg would attract purchasing power that locals could use. The owner of the kumpir shop seems to particularly hire local young women of colour, including Muslim women who wear hijab. It is for this reason that Yenirce asks, "who is gentrifying whom here exactly?" (2017a, translation by author). Is it racialized Wilhelmsburgers who have built businesses to meet the demands of white, German New Wilhelmsburgers? Or the white, German New Wilhelmsburgers who have arrived on the island in the past decade, some of them with an analysis of the process in which they play a role?

Racialized Wilhelmsburgers' accounts of recent changes in Wilhelmsburg suggest that it is a minority who are in a position to take much advantage of opportunities that the "mixing" of the neighbourhood presents. Many more are affected by the housing situation that residents are concerned about. I see this as the crux of the complex and conflictual picture of Wilhelmsburgers' perspectives as I have laid them out in this chapter: they have hopes and aspirations for what mixing can offer Wilhelmsburg, but based on residents' observations so far, these hopes have not yet been realized. Perhaps it is still early in the change process; my research began just a few years after the end of the IBA, and as one local put it to me, the full effects of the policies and projects that I described in the previous chapter will only be visible once "the middle-class armada" has truly landed (personal communication, 2018). But as welcome as New Wilhelmsburgers – white, German, middle class students and families – are to the island, and as hopeful as many residents are about the benefits, some racialized residents simply feel that the changes happening in the neighbourhood are not *for* them.

> It has advantages and disadvantages, I would say. The advantage is that the neighbourhood is experiencing a change, that people with prospects live here. On the other hand, it always had its own charm, and that is being stripped away... We enjoy for example that there are restaurants, that's something we find nice, that there is a park, that there are the Towers [Hamburger basketball team now located in Wilhelmsburg], that it is becoming a bit livelier. But I find it sad that it wasn't done for us. That's what I see when I look at it: The change is not happening for the people who live here but for those who have come in the last two, three years. We have lived with zero support for 30, 40 years and it also isn't support to do nothing for 30, 40 years and then remove the people who are sort of at the bottom by not offering them apartments, or the prices go up so much that they can't live here anymore. That just means you transfer the problem to another neighbourhood (Arzu, interview, 2017).

Arzu argues that the city should have supported the neighbourhood from within, supported the people already living on the island to "improve" it. As it is, she worries that its charm is just being destroyed and its fundamental problems left unsolved. Yücel also concluded that the changes that he could see in the neighbourhood had not made a dent in the problems with which he is really concerned, the problems of poverty and limited life chances for Wilhelmsburg's young population:

> It looks really good, the houses, the apartments, the new [footbridge by S-Bahn Wilhelmsburg], and as a result it has gotten a lot more expensive. Wilhelmsburg has gotten expensive, and otherwise I don't notice anything, nothing to really reach the disadvantaged and help, there was nothing there ... nothing for our lives, our social disadvantage (Yücel, interview, 2017).

As Arzu (interview, 2017) put it: "the people who have little are still here, still need support," yet that support has not materialized through the mechanisms of mix that residents generally support. Instead, alongside new arrivals, new services and changes to the built environment, the landscape in Wilhelmsburg has gotten harsher for people whose access to resources is limited by structural positioning within a racialized economic system.

It remains an open question whether residents' hopes about the promise of mixing will in some way be reconciled with the processes that they observe that do not serve the racialized and low-income population of Wilhelmsburg. Residents to whom I presented my preliminary findings were concerned that mix may undermine the familiarity and solidarity that has historically characterized the island, damaging Wilhelmsburgers' capacity to band together to make planning of the island work for them. There is also the potential for non-hegemonic aspects of residents' hopes for social mix to simply be absorbed into the hegemonic discourse, in the way that Haritaworn (2012) demonstrates mixed subjects are continuously appropriated through commodification and fetishization. The rearticulation of social mix as an anti-racist intervention could further serve dominant interests by allowing white Germans, New Wilhelmsburgers more specifically, to feel good about themselves (see Ahmed, 2004, para. 34) while consuming the "diversity" of the people and neighbourhood that is being mixed through their arrival (Ahmed, 2000; hooks, 1992). I have discussed elsewhere (Chamberlain, 2012, 2013) how IBA Hamburg's framing of the neighbourhood encouraged precisely that. There is ample evidence that such a process is not particularly transformative of the fundamental power relations that have shaped the neighbourhood, and with which racialized Wilhelmsburgers are deeply concerned (cf. Ahmed, 2000, 2004; Haritaworn, 2012; hooks, 1992).

In the desire to live together across difference, however, I see an opportunity as well as an area for future research, as I will describe in the Conclusion. Wilhelmsburgers' appetite for connection and relationship, for a conviviality that is deep and substantial, could provide a basis for strategic organizing beyond the individualized, state-structured participation opportunities. Old Wilhelmsburgers know their neighbourhood, and with a platform their knowledge might be mobilized to advocate for their interests.

Conclusion

This chapter focused on racialized residents' perspectives on recent developments in Wilhelmsburg, starting from my questions about what changes, if any, had been taking place in the neighbourhood. I drew on my interviews to describe the arrival of new people to Wilhelmsburg and the increased "mix" in the neighbourhood. I demonstrated that figures of "Bulgarians" and "Germans" were central to how long-time residents described these arrivals. Wilhelmsburgers argued that the arrival of white, middle-class Germans represented a change for the island, while an influx of Bulgarians, Romanians, and other Eastern Europeans did not, as the latter represents a continuation of Wilhelmsburg as an "Arrival City" and immigrant neighbourhood. The residents I spoke to tied "mix" to what they saw as positive and negative developments on the island, and in so doing talked about mixing in ways that both reproduced and challenged the common sense notion. I argued that there is a fundamental tension wherein racialized residents observed and critiqued the effects of mixing policies yet sometimes also supported the logic of the policies. This presents a challenge for residents in the context of the changing neighbourhood, and is an area in which this research makes a contribution to the community, as a tool for discussion and reflection, as I will discuss in the Conclusion that follows.

In the second half of the chapter I brought my interviews together with participant sensing and archival research to look at themes of conflict and tension between different residents of Wilhelmsburg, in order to illustrate how residents themselves recognize that "mix" does not necessarily equate to the kind of *living together* that they hope for, particularly with white Germans, given the historical stigmatization and isolation of the island. The long-time residents I interviewed reproduced "Bulgarians" as a distinct and deviant group through discourses of fear that they questioned and mobilized. In relation to "Germans" I outlined accounts of distance and conflicting interests and analyses of differential power relations in the neighbourhood. The question of how to live well together across difference in this context was one that many residents wanted to talk about. I see potential for this discussion in connection with urgent advocacy for the needs of Wilhelmsburg's most vulnerable, and indeed for all who want to stay in the neighbourhood in the long term. In the following Conclusion, I will reflect on what residents, local planners and politicians, and urban researchers might take away, looking back at what the book has accomplished, and forward to a research program for the future.

Conclusion

Introduction

In this study, I asked about the experiences of racialized residents of Hamburg-Wilhelmsburg with the developments that have recently taken place in their neighbourhood, and about how racialized people have figured into planning on the island. I came to this project through a distinct gap in urban research in Germany, where the salience of racism and racialization to planning and the urban landscape has been largely ignored, or obscured. I followed the lead of Black German and German scholars of Colour who have argued for urban research that takes racism and colonialism seriously, as structures and processes that shape cities in Europe and in Germany. Once I began the research, I also followed the direction of the residents I interviewed, to frame the snapshot of Wilhelmsburg in the midst of a process of gentrification that is distinctly racialized, as planning of the neighbourhood has been for over a century. The insights that this approach yielded were rich, and broader than the research questions I started with. In this conclusion, I characterize what I see as the contributions of this research and this book, beginning with a summary of the ground that it covered and my thoughts on the significance to urban scholarship, planning, and to the community. I conclude with directions for future research.

Research Contributions

I began, in Chapter 1, by presenting the theoretical framework that has guided my approach to this research and my analysis. First, I defined racialization and racism as processes and structures that are embedded in the production of German identity and citizenship, in law, and in urban planning. At the crux of locally specific racialization is the externalization of racialized "others" as un-German, and the migrantization of people who are in fact at home in the country.

I further rooted the dissertation in theories of how racialization and racism as prevailing societal structures and processes are spatialized in and through cities in

Figure 49: "We are here to stay" reads graffiti on the steps up the dike, spotted from across Harburger Chaussee

(photo by author, 2017)

forms and state strategies that vary, but in which the devaluation and control of racialized people is consistent. The racialization of space, segregation, and ghettoization are crucial concepts that I used alongside the concept of racialized displaceability to analyze past and present developments in Hamburg-Wilhelmsburg. I particularly focused in on social mix as a form of racial control and domination that is tremendously influential and common sense in German urban planning. I suggested a reading of this spatialized and racialized devaluation through the concept of racial capitalism, in which racialized devaluation of people and of land are understood as normal and integral parts of the economic system.

In Chapter 2, I outlined my approach to the study in Hamburg-Wilhelmsburg. I situated the study as an anti-racist, intersectional public ethnography and described how anti-racist and intersectional methodologies informed the research questions and the structure of the study. I centred myself in this chapter, explaining how I came to do my doctoral research in and about Wilhelmsburg, and I located myself within an understanding of ethnography as embodied and self-reflexive. I described my methods of data collection: ethnographic interviews with racialized residents and local politicians and planners, participant sensing, researcher-driven photography, and archival work. I also described how and why I complemented these methods with an interim report to participants and a community event that used visual and textual tools to prompt discussion of my preliminary findings. These are some of the ways in which I attempted to operationalize anti-racist methodology as relational and reciprocal, and to manage the challenges inherent to the project.

I then presented four data-centred chapters, which were organized according to the analysis and priorities that I heard from interviews with racialized residents of Wilhelmsburg. I asked the reader to follow a slightly unconventional structure as a result, in which I framed residents' views of the island as the context in which to understand urban development and planning. Residents encouraged me to look at the long history of Wilhelmsburg as an immigrant neighbourhood to understand the historical development of the island and to understand the significance of the "bad reputation" that presaged the recent push to revalue and redevelop the neighbourhood. This was the focus of Chapter 3, in which I drew on interviews and government and media archives to fill in the blanks and silences in the dominant planning narrative of how and why Wilhelmsburg required intervention. I situated the island's history as a racialized space planned for work and waste within the long-standing structures of racial capitalism and within Hamburg as city shaped and enriched by colonialism. I followed the insights of Black and People of Colour scholars and activists in Hamburg who have argued that racism and colonial exploitation have made Hamburg what it is today, and have particularly defined its relationship to its "others" and "other spaces."

In Chapter 4 I turned to a prominent theme that emerged from my interviews with residents: the valuation of the island not as a "problem neighbourhood" or a "ghetto," but as a beautiful and warm place to live, particularly in the context of societal racism. I emphasized how Wilhelmsburgers called the island *Heimat*, a space of identity and belonging. In contrast to the often-racist uses of the concept, Wilhelmsburgers framed this *Heimat* as a space of conviviality and multiculture, where racialized people who are excluded or considered suspect in other neighbourhoods, belong and are represented. I demonstrated how this reading of the neighbourhood resists the dominant discourse about it, while also resisting the devaluation of racialized people. Situating Wilhelmsburg as *Heimat* articulates grounds of belonging *as racialized people and immigrants* in a country where national identity and belonging is exclusive. I see this as a practice of hope and of self-determination that can be learned from and built upon.

I argued, however, that the urban planning interventions recently implemented by Hamburg city-state represent a threat to Wilhelmsburg as racialized home in the long term. As I indicated in the Introduction, I chose to explore Wilhelmsburg as *Heimat* first, as crucial context for an analysis of recent development on the island. In Chapter 5, I discussed the IBA Hamburg and International Garden Show event-projects, the subsidy for student housing, and social mix policy in public housing as instruments to encourage the settlement of more white, middle-class Germans on the island, and thus social mixing of the neighbourhood as a whole. While local researchers and activists flagged this as a process of gentrification that puts low-income residents at risk of exclusionary displacement over time, I demonstrated that what is underway is a specifically racialized mixing strategy. The strategy is anchored in the white supremacy that is enshrined and protected in Germany's so-called anti-discrimination law, which in fact allows discrimination in the name of social mixing.

I have strongly questioned whether the mixing of Wilhelmsburg is compatible with the maintenance of the island as *Heimat* in the context of exclusion. The evidence on social mix as an instrument that breaks up communities, undermines solidarity, and reduces the ability of long-term residents to stay in the neighbourhood, suggests that social mixing here will likely be detrimental to many of the neighbourhood qualities that Wilhelmsburgers deeply value. As Arzu put it, Wilhelmsburg "has always had its own charm, and that's being stripped away." Yet in Chapter 6, I have identified real complexity and contradiction in racialized Wilhelmsburgers' perspectives on what is happening in the neighbourhood so far. On the one hand, residents observed with some enthusiasm that white Germans have begun moving to Wilhelmsburg in greater numbers. They saw this movement as an end to the long-standing stigmatization of the neighbourhood, and as an opportunity to live more closely with white Germans and to bridge the divides they have experienced for a long time. In this sense, residents viewed social mix differently than both the dominant logic and my reading of it as an outsider, and imbued it with hope and anti-racist potential. On the other hand, I also demonstrated that Wilhelmsburgers perceived very troubling developments in the neighbourhood, particularly the impact on the cost and availability of housing, that I view as tied to the policies and processes of social mixing. I identified a number of areas of tension and conflict among Wilhelmsburgers who are differently located in relation to social, economic, and political power to show that residents themselves saw conviviality as already conflictual and challenging. The book thus leaves a number of open questions, including whether and how the promise and negative impact of social mixing can or should be reconciled from the perspectives of Wilhelmsburg residents. What will happen next? Might the anti-racist potential of mixing be realized? How?

My hope is that, among its contributions, this research will support Wilhelmsburgers' self-reflection and consideration of the effects of social mixing on the island that they love. The book reflects Wilhelmsburgers' perspectives back to them and particularly amplifies the voices of racialized residents that have, to date, not been front and centre in planning discourse. My exploration of residents' use of the concept of *Heimat* also connects residents with the ongoing public debate about the concept in Germany. The book demonstrates that racialized Wilhelmsburgers are understanding and imagining *Heimat* in ways that fundamentally unsettle its racist history and serve to articulate community belonging and connection. This is, as Nalan noted (see p. 161), a strength that deserves recognition. I hope that the research can thus play a role in resident advocacy and activism, and in local strategizing to address the challenges facing racialized island residents in the coming years.

This book also contributes to the existing body of knowledge about the social mix common sense in German urban planning and beyond. Residents' complex articulations of social mixing as a potentially anti-racist strategy are, to my knowledge, unique in the literature and deserve further attention as non-hegemonic recasting of a hegemonic discourse, despite the tensions and cautions I have highlighted. In contrast, my analysis of social mixing as a specifically racial logic, and as grounded in white supremacy that is enshrined in German law, provides a sharp, race-critical argument against the common sense. This is the primary contribution of the book to urban planning practice and discourse in Hamburg and beyond: the message that a poorly supported, racially articulated 'myth' should be untenable to planners and politicians who are invested in social and urban justice. To this extent the book is also intended to be an instrument of self-reflection for this audience as well, and another step towards the tipping point in transforming planning common sense.

With this book I hope to add to the small body of critical race urban research on the German context. By applying concepts that are not common in German urban research, particularly that of racial capitalism and environmental racism, I have centred racialization in my analysis of the neighbourhood and the city. This enables, as Roy (2018) identified, the pinpointing of the role of difference constructed in colonialism in urban political economy, a role that has been largely obscured to date by scholarly silence around race, the migrantization of urban "others," and thus the imagination of the urban subject as always white (see Ha & Picker, 2022, for discussions of this in the European urban context). My hope is that this book aids in the normalization of anti-racist urban analysis in and about Germany, and to the undermining of the notion that people who talk about racism are themselves "the problem." This thinking has seriously limited the capacity of German researchers to grapple with the realities and challenges of urban life for all city dwellers. This research contributes to the prospects for urban justice by modelling what it can mean to take racialization seriously in scholarly work, at a moment in which public activism against racism and the simultaneous deepening and illumination of structural inequalities during the pandemic underscore the urgency of this imperative.

Directions for Future Research

The research also illuminated some key areas for future research and action. One of the questions that I frequently encountered among the Wilhelmsburgers I interviewed, as well as those who supported me through the research process, was how Wilhelmsburg residents could live more closely together across differences: living more *with one another,* rather than *alongside one another.* This was not a question that I was able to address in this study, but it was closely tied to the themes and realities of social mixing as it is playing out on the island. One of the great strengths of *Heimat* Wilhelmsburg is that it does not depend on harmony or inflexibility: it can accommodate newcomers and cope with tension. In my interviews, several of the racialized residents I spoke to asked if I would interview some recently arrived white Germans as well, as they were interested to hear *their* perspectives about the island. I also met progressive white German community organizers who were wondering about how they could connect with Wilhelmsburgers of various linguistic communities and migration statuses to work on neighbourhood issues. I think the mutual interest in connection and engagement could present an opportunity for action research to explore what it means, in fact, to live "with one another" in a changing neighbourhood, in a way in which power differentials and systemic inequality are acknowledged and perhaps exploited in the service of anti-racist goals. This book has demonstrated that any future research needs to take racism seriously as it has shaped urban development in Hamburg and in Wilhelmsburg to date.

Likewise, social mixing must be approached in future research as a racialized planning strategy. There is scope for a closer look at how and why this racialized technique continues to be a planning common sense in Germany when there is now a wealth of evidence that it can harm the communities where it is implemented. Wilhelmsburg might also be an appropriate case study of a neighbourhood where social mix planning has been implemented. Talking to long-time residents again in five and ten years would provide an opportunity to assess what happens to racialized residents over time, from their own perspectives. Further, the question of what happens to environmental conditions on the island is a pressing one. Are environmental threats and burdens eliminated from the neighbourhood as it is revaluated? What does mixing do to conditions of environmental racism?

The third area for future research that emerged from this study is an exploration of how notions of planning and management of urban diversity travel internationally, in the context of global structures of racial capitalism and colonialism. During my research in Hamburg, I was often confronted with questions about what Hamburg and other German cities could learn from my hometown of Toronto, as a city where diversity is apparently considered "our strength." At the same time, the planners and politicians who I spoke to often had experiences in Toronto or with Torontonians that focused on policy transfer particularly related to urban diversity.

There tends to be silence in these exchanges when it comes to Toronto as a settler colonial city, where the diversity framework contributes to the settler colonial logic by attempting to make the Indigenous presence invisible and reducing it to just one "diversity" among many. Research into policy transfer between Toronto, Hamburg, and other German cities would build upon this study's findings about the central logic of social mix and put Wilhelmsburgers' experiences into conversation with those of residents in other cities.

Finally, this research contributes to the case currently being made by German People of Colour and Black Germans against "migration background" as a statistical instrument, and for data collection that would actually help to illuminate and document discrimination in Germany (Ahyoud et al., 2018; neue deutsche organisationen, 2017). The statistical and commonplace concept of migration background has appeared throughout this book as a powerful and problematic discourse that externalizes and migrantizes certain Germans while failing to meaningfully reflect their experiences. Social mix as it is practiced in Wilhelmsburg and beyond is likely also not the only planning common sense that is enabled by the racializing notion of migration background and by the lack of data on discrimination (Münch, 2009). This book has demonstrated that whether planners and politicians pay attention to it or not, racial discrimination is playing a significant role in racialized people's experiences of urban life and belonging. It is time for urban research in Germany to catch up with what racialized Germans have been clearly indicating for decades and to produce evidence that communities can use for action and advocacy in service of a vision of well being beyond the limitations of current systems and structures.

Bibliography

Adanalı, Z. (2013). Türkischstämmige Migranten auf dem Hamburger Wohnungsmarkt: Am Beispiel des Stadtteils Wilhelmsburg [Masters Thesis]. HafenCity Universität.

Afrika-Hamburg.de. (n.d.). Retrieved February 9, 2019, from https://www.afrika-hamburg.de/index.php

Ahmed, S. (2000). Strange encounters: Embodied others in post-coloniality. Routledge.

Ahmed, S. (2002). Racialized bodies. In M. Evans & E. Lee (Eds.), Real bodies: A sociological introduction (pp. 46–63). Palgrave Macmillan.

Ahmed, S. (2004). Declarations of whiteness: The non-performativity of anti-racism. Borderlands, 3(2). https://www.borderlands.net.au/vol3no2_2004/ahmed_declarations.htm

Ahmed, S. (2013, September 11). Making Feminist Points. Feministkilljoys. https://feministkilljoys.com/2013/09/11/making-feminist-points/

Ahyoud, N., Aikins, J. K., Bartsch, S., Bechert, N., Gyamerah, D., & Wagner, L. (2018). Wer nicht gezählt wird, zählt nicht: Antidiskriminierungs- und Gleichstellungsdaten in der Einwanderungsgesellschaft – Eine anwendungsorientierte Einführung. Vielfalt entscheidet - Diversity in Leadership, Citizens for Europe.

Aikins, M. A., Bremberger, T., Aikins, J. K., Gyamerah, D., & Yıldırım-Caliman, D. (2021). Afrozensus 2020: Perspektiven, Anti-Schwarze Rassismuserfahrungen und Engagement Schwarzer, afrikanisher und afrodiasporischer Menschen in Deutschland. https://afrozensus.de

Allen, T. W. (2012). The invention of the white race (Second Edition). Verso.

Amnesty International. (2016). Living in insecurity: How Germany is failing victims of racist violence. Amnesty International.

Analyse & Konzepte. (2013). IBA Strukturmonitoring 2013: Endbericht. IBA Hamburg GmbH. https://www.iba-hamburg.de/wissen/strukturmonitoring.html

Andersen, H. S. (2017). Selective moving behaviour in ethnic neighbourhoods: White flight, white avoidance, ethnic attraction or ethnic retention? Housing Studies, 32(3), 296–318.

Anderson, K. (1991). Vancouver's Chinatown: Racial discourse in Canada, 1875-1980. McGill-Queen's University Press.

Anthias, F., & Yuval-Davis, N. (1989). Introduction. In Woman – Nation – State (pp. 1–15). St. Martin's Press.

Arani, A. Y. (2015). Anti-Muslim Racism and Islamophobia in Germany: Alternative Report on the 19th – 22nd State Report of the Federal Republic of Germany in accordance with Article 9 the International Convention on the Elimination of All Forms of Racial Discrimination (ICERD) (p. 59). The Network Against Discrimination and Islamophobia, Inssan e.V.

Araujo, M. (2016). A very "prudent integration": White flight, school segregation and the depoliticization of (anti-) racism. Race, Ethnicity and Education, 19(2), 300–323.

Arbeitskreis Hamburg Postkolonial. (n.d.). Arbeitskreis HAMBURG POSTKOLONIAL willkommen. Retrieved November 23, 2018, from https://www.hamburg-postkolonial.de/willkommen.html

Arbeitskreis Hamburg Postkolonial. (2018, April 29). Die Hafencity zwischen kosmopolitischem Flair und kolonialen Echos [Walking Tour].

Arbeitskreis Umstrukturierung Wilhelmsburg. (2011). Alles, alles verkehrt: Wie das Stadtentwicklungsinstrument "Internationale Bauausstellung" Wilhelmsburg und die Veddel verwertbar aufbereiten soll und was das für die BewohnerInnen bedeutet. https://www.blog.de/community/profile_documents.php?user=AKU-Wilhelmsburg

Arbeitskreis Umstrukturierung Wilhelmsburg. (2013a). Unternehmen Wilhelmsburg: Stadtentwicklung im Zeichen von IBA und igs. Assoziation A.

Arbeitskreis Umstrukturierung Wilhelmsburg. (2013b, July 22). Offener Brief an Saskia Sassen. Arbeitskreis Umstrukturierung Wilhelmsburg. https://akuwilhelmsburg.blogsport.eu/2013-07-offener-brief-an-saskia-sassen/

Ardal, S., Bababoutilabo, V., & Perinelli, M. (n.d.). Rassismus und Corona II (No. 2). Retrieved April 3, 2020, from https://www.rosalux.de/publikationen/mediathek/manypod

Arslan, E. (2015). Turkish Berlin: Integration policy and urban space. Ethnic and Racial Studies, 38(3), 460–462.

Arthurson, K. (2012). Social mix and the city: Challenging the mixed communities consensus in housing and urban planning policies. Csiro Publishing.

Ataman, F. (2018a, June 12). Deutschland, Heimat der Weltoffenheit. Amadeu Antonio Stiftung. https://www.amadeu-antonio-stiftung.de/deutschland-heimat-der-weltoffenheit-8409/

Ataman, F. (2018b, June 16). Integration unter Seehofer: Das ist auch unsere Heimat, Herr Heimatminister! Spiegel Online. https://www.spiegel.de/kultur/gesellschaft/horst-seehofer-das-ist-auch-unsere-heimat-herr-heimatminister-kolumne-a-1213241.html

Attia, I., & Popal, M. (2016). Antimuslimischer Rassismus dekolonial: Kontrapunktische Lektüren westlicher Islamdiskurse. Das Argument, 319, 651–660.

Aydemir, F., & Yaghoobifarah, H. (2019). Vorwort. In Eure Heimat ist unser Albtraum. Ullstein Buchverlage.

Bacchetta, P., El-Tayeb, F., & Haritaworn, J. (2015). Queer of colour formations and translocal spaces in Europe. Environment and Planning D: Society and Space, 33(5), 769–778.

Bailey, C. A. (2013). Public ethnography. In S. N. Hesse-Biber & P. Leavy (Eds.), Handbook of emergent methods (pp. 265–281). The Guildford Press.

BAMF. (2017). Glossar. Bundesamt Für Migration Und Flüchtlinge. https://www.bamf.de/DE/Service/Left/Glossary/_function/glossar.html?lv3=3198544

Barman, J. (2005). Stanley Park's secret: The forgotten families of Whoi Whoi, Kanaka Ranch and Brockton Point. Harbour Publishing.

Barskanmaz, C. (2011). Rasse – Unwort des Antidiskriminierungsrechts? Kritische Justiz, 3, 382–389.

Barskanmaz, C. (2012, June 16). Critical race theory beyond German exceptionalism. Critical Race Theory Europe Symposium, Humboldt University, Berlin. https://vimeo.com/66955298

Barskanmaz, C., & Samour, N. (2020). Debatte – Falsches Signal. Der Freitag. https://www.freitag.de/autoren/der-freitag/falsches-signal

Barth, A. (2000, October 30). STÄDTE: „Ein ungeheuer belastendes Klima". Der Spiegel, 44. https://www.spiegel.de/spiegel/print/d-17704702.html

Baumgartner, N. (2019, May 27). Flüchtlingsheim Hafencity: Container auf teurem Baugrund. FINK.HAMBURG. https://fink.hamburg/2019/05/fluechtlingsheim-hafencity-container-auf-teurem-baugrund/

Bayer, N., Holm, A., & Lebuhn, H. (2014). Städtische Diskurse um Migration im Wandel. Integration, diversity und soziale Bewegungen in München und Berlin. sub\urban. zeitschrift für kritische stadtforschung, 2(3), 81–92.

Behörde für Kultur und Medien. (n.d.). Erinnerungskultur: Aufarbeitung des kolonialen Erbes. Hamburg.de. Retrieved February 8, 2019, from https://www.hamburg.de/bkm/koloniales-erbe/

Behörde für Kultur und Medien. (2021). Workshop Nr. 1 zur Neukontextualisierung des Bismarck-Denkmals. hamburg.de. https://www.hamburg.de/bkm/koloniales-erbe/15177680/bismarck-denkmal-workshop-1/

Behörde für Schule und Berufsbildung. (n.d.). Harburg/Wilhelmsburg: Stadtteile Harburg und Wilhelmsburg in der NS-Zeit. Hamburg.de. Retrieved February 12, 2019, from https://www.hamburg.de/clp/dabeigewesene-begriffserklaerungen/clp1/ns-dabeigewesene/onepage.php?BIOID=782

Behörde für Stadtentwicklung und Umwelt. (2005). Sprung über die Elbe: Hamburg auf dem Weg zur internationalen Bauausstellung – IBA Hamburg 2013.

Freie und Hansestadt Hamburg; https://www.iba-hamburg.de/fileadmin/Die_IBA-Story_post2013/051030_sprung_ueber_die_elbe.pdf.

Behörde für Stadtentwicklung und Umwelt. (2013, Oktober). Neuer Park für Wilhelmsburg: Igs 2013 schließt ihre Tore. Hamburg.de. https://www.hamburg.de/pressearchiv-fhh/4119688/2013-10-13-bsu-igs/

Behörde für Stadtentwicklung und Umwelt, HafenCity University, & University of Hamburg. (2010). Pilotbericht 2010 im Sozialmonitoring im Rahmenprogramm Integrierte Stadtteilentwicklung. Freie und Hansestadt Hamburg. https://www.hamburg.de/sozialmonitoring/4603308/rise-pilotbericht-2010/

Behörde für Stadtentwicklung und Wohnen. (n.d.). Förderprogramm: Wohnen für Studierende und Auszubildende. hamburg.de. Retrieved November 14, 2018, from https://www.hamburg.de/mieten/4049334/wohnen-studierende-auszubildende/

Bezirksamt Hamburg-Mitte. (2015). Sozialraumbeschreibung Wilhelmsburg. Bezirksamt Hamburg-Mitte: Fachamt Sozialraummanagement.

Binde, N. (2019, February 20). Kreative legen neues Konzept für Zinnwerke vor. https://www.abendblatt.de/hamburg/article216484683/Kreative-legen-neues-Konzept-fuer-Zinnwerke-vor.html

Birke, P. (2013). Radikaler Umbau: Die Internationale Bauausstellung in Hamburg-Wilhelmsburg als Labor der neoliberalen Stadtentwicklung. Emanzipation, 3(2), 98–109. https://www.emanzipation.org/?q=ema_3-2.

Black Lives Matter Berlin. (2020, June 6). Stellungnahme zu rassistischer Polizeigewalt. Black Lives Matter Berlin. https://www.blacklivesmatterberlin.de/stellungnahme-zu-rassistischer-polizeigewalt/

Blickle, P. (2002). Heimat: A critical theory of the German idea of homeland. Camden House.

Blokland, T., & Nast, J. (2014). From public familiarity to comfort zone: The relevance of absent ties for belonging in Berlin's mixed neighbourhoods. International Journal of Urban and Regional Research, 38(4), 1142-11 59.

Boa, E., & Palfreyman, R. (2000). Heimat: A German Dream. Oxford University Press.

Boccagni, P. (2016). Migration and the search for home: Mapping domestic space in migrants' everyday lives. Palgrave Macmillan.

Bolt, G. (2009). Combating residential segregation of ethnic minorities in European cities. Journal of Housing and the Built Environment, 24(4), 397-405.

Bolt, G., Phillips, D., & Kempen, R. V. (2010). Housing Policy, (De)segregation and Social Mixing: An International Perspective. Housing Studies, 25(2), 129–135.

Bråmå, Å. (2006). "White Flight"? The production and reproduction of immigrant concentration areas in Swedish cities, 1990-2000. Urban Studies, 43(7), 1127–1146.

Bridge, G., Butler, T., & Lees, L. (2012). Introduction: Gentrification, social mix/ing and mixed communities. In Mixed communities: Gentrification by stealth (pp. 1–16). Policy Press.

Brinkbäumer, K. (2000, July 3). Er machte alle kalt. Der Spiegel, 27. https://www.spiegel.de/spiegel/print/d-16810637.html

Brosda, C. (2018, April 6). Address by Senator Dr Carsten Brosda. Senate Reception for the Herero and Nama Delegation of the Second Transnational Herero and Nama Congress, Rathaus, Kaisersaal, Hamburg. https://www.hamburg.de/contentblob/10903692/f7b049eb4294c0315ce65bf12ed2c5e5/data/rede-brosda-se-herero-nama.pdf

Brück, D., & David, I. (2014, 05). Inselpark in Wilhelmsburg: Der Zaun bleibt. Mittendrin: Das Nachrichtenmagazin für Hamburg-Mitte. https://hh-mittendrin.de/2014/05/inselpark-in-wilhelmsburg-der-zaun-bleibt/

Bucerius, S. (2012). "What Do You Expect? That We All Dance and Be Happy?" Second-Generation Immigrants and Germany's 1999 Citizenship Reform. German Politics & Society, 30(1 [102]), 71–85.

Bucerius, S. (2014). Unwanted: Muslim immigrants, dignity and drug dealing. Oxford University Press.

Bullard, R. D. (1996). Environmental justice: It's more than waste facility siting. Social Science Quarterly, 77(3), 493–499.

Bullard, R. D. (2018). Dumping in Dixie: Race, class, and environmental quality (Third Edition). Routledge.

Bundesministerium der Justiz und für Verbraucherschutz. (2006). Allgemeines Gleichbehandlungsgesetz. https://www.gesetze-im-internet.de/englisch_agg/englisch_agg.html#p0101

Bundesministerium der Justiz und für Verbraucherschutz. (2017). BauGB - Baugesetzbuch. https://www.gesetze-im-internet.de/bbaug/BJNR003410960.html

Bürgerhaus Wilhelmsburg. (2018). Perspektiven! Miteinander planen für die Elbinsel. Bürgerhaus Wilhelmsburg. https://buewi.de/aktivitaeten/perspektiven/start

Bürgerschaft der Freien und Hansestadt Hamburg. 19 Walhperiode. (2009). Schriftliche Kleine Anfrage des Abgeordneten Metin Hakverdi (SPD) vom 13. 11. 09 un Antwort des Senats. Betr.: Wie viel geben IBA und igs für Werbung und Öffentlichkeitsarbeit aus? Freie und Hansestadt Hamburg.

Bürgerschaft der Freien und Hansestadt Hamburg, 22. Wahlperiode. (2021). Mitteilung des Senats an die Bürgerschaft: Stellungnahme des Senats zu dem Ersuchen der Bürgerschaft vom 13. Januar 2021 „Umfassende Überprüfung der Freistellungsgebiete" (Drucksache 22/2676). Freie und Hansestadt Hamburg.

Byrne, J., & Wolch, J. (2009). Nature, race, and parks: Past research and future directions for geographic research. Progress in Human Geography, 33(6), 743–765.

Çağlar, A. (1995). German Turks in Berlin: Social exclusion and strategies for social mobility. Journal of Ethnic and Migration Studies, 21(3), 309–323.

Çağlar, A. (2001). Constraining metaphors and the transnationalisation of spaces in Berlin. Journal of Ethnic and Migration Studies, 27(4), 601–613.

Çağlar, A., & Schiller, N. G. (2018). Migrants and city-making: Dispossession, displacement, and urban regeneration. Duke University Press.

Çalışkan, G. (2011). Forging diasporic citizenship: Berlin's German-born Turkish Ausländer [Doctoral Dissertation]. York University.

Castañeda, E. (2018). A place to call home: Immigrant exclusion and urban belonging in New York, Paris, and Barcelona. Stanford University Press.

Chamberlain, J. (2012). Who is Wilhelmsburg? Race and space in Internationale Bauausstellung Hamburg [Masters Thesis]. University of Toronto.

Chamberlain, J. (2013). Problem place, problem people: Spatialized racial discourses in an urban planning project in Hamburg, Germany. CES Contexto, 2, 780–792.

Chamberlain, J. (2020). Experimenting on racialized neighbourhoods: IBA Hamburg and the urban laboratory in Hamburg-Wilhelmsburg. Environment and Planning D: Society and Space. https://doi.org/10.1177/0263775820903328

Chamberlain, J. (2022). Theorizing Hamburg from the South: Racialisation and the development of Wilhelmsburg. In G. Picker & N. Ha (Eds.), European cities: Modernity, Race, and Colonialism (pp. 235–256). Manchester University Press.

Chamberlain, J., August, M., Rinn, M., & Ruiz-Tagle, J. (2022, May 8). Racism and Social Mix. On Kip, M. & Beveridge, R. (Podcasters). Urban Political Podcast. https://urbanpolitical.podigee.io/51-social_mix_racism

Cheshire, P., Gibbons, S., & Gordon, I. (2008). Policies for "Mixed Communities:" A Critical Evaluation [SERC Policy Paper]. Spatial Economics Research Centre.

Chin, R. (2007). The guestworker question in postwar Germany. Cambridge University Press.

Chin, R. (2009). Guestworker migration and the unexpected return of race. In R. Chin, H. Fehrenbach, G. Eley, & A. Grossman (Eds.), After the Nazi racial state: Difference and democracy in Germany and Europe (pp. 80–101). University of Michigan Press.

Chin, R., Fehrenbach, H., Eley, G., & Grossman, A. (2009). After the Nazi racial state: Difference and democracy in Germany and Europe. University of Michigan Press.

Christmann, G. B. (2013). Raumpioniere als Gestalter in sozial benachteiligten Stadtquartieren? Über Inklusions- und Exlusionsprozesse in der Quartiersentwicklung "von unten." In K. Harm & J. Aderhold (Eds.), Die subjektive Seite der Stadt: Neue politische Herausforderungen und die Bedeutung von Eliten im lokalen Bereich (pp. 97–118). Springer-Verlag.

Christmann, G. B. (2014). Investigating spatial transformation processes. An ethnographic discourse analysis in disadvantaged neighbourhoods. Historical Social Research / Historische Sozialforschung, 39(2 (148)), 235–256.

Clausen, S. (2008). Süße Heimat Wilhelmsburg: Neuer Blick auf die lokale Geschichtsschreibung. In Geschichtswerkstatt Wilhelmsburg Honigfabrik e.V. & Museum Elbinsel Wilhelmsburg e. V. (Eds.), Wilhelmsburg: Hamburgs große Elbinsel (pp. 175–189). Medien-Verlag Schubert.

Colonial Amnesia: Quo Vadis, Hamburg? (n.d.). YouTube. Retrieved June 11, 2018, from https://www.youtube.com/playlist?list=PLmnAZlzZZovKp-RxC9BVbOG59VXPEscos

Connolly, K., & Oltermann, p. (2020, February 20). Hanau attack reveals "poison" of racism in Germany, says Merkel. The Guardian. https://www.theguardian.com/world/2020/feb/20/hanau-attack-reveals-poison-of-racism-in-germany-says-merkel

Copwatch HH. (n.d.). Stop racist police controls! – Gerade in Zeiten von Corona. Copwatch-Hamburg. Retrieved July 11, 2020, from https://copwatchhamburg.blackblogs.org/en/stop-racist-police-controls-especially-in-times-of-corona-copwatchhh/

Costadura, E., Ries, K., & Wiesenfeldt, C. (2019). Heimat global: Einleitung. In Heimat global: Modelle, Praxen und Medien der Heimatkonstruktion (pp. 11–42). transcript Verlag.

Crenshaw, K. (1991). Mapping the margins: Intersectionality, identity politics, and violence against women of color. Stanford Law Review, 43, 1241–1299.

Danewid, I. (2019). The fire this time: Grenfell, racial capitalism and the urbanisation of empire. European Journal of International Relations, 1–25.

David, I. (2015, April 20). Wilhelmsburg: Von der Gartenschau zum Inselpark für alle Menschen. Mittendrin: Das Nachrichtenmagazin für Hamburg-Mitte. https://hh-mittendrin.de/2015/04/wilhelmsburg-von-der-gartenschau-zum-inselpark-fuer-alle-menschen/

Davidson, M. (2008). Spoiled mixture: Where does state-led 'positive' gentrification end? Urban Studies, 45(12), 2385–2405.

Davoli, M., & Entorf, H. (2018). The PISA shock, socioeconomic inequality, and school reforms in Germany (IZA Policy Paper No. 140). IZA – Institute of Labor Economics.

de Koning, A. (2015). "This neighbourhood deserves an espresso bar too:" Neoliberalism, racialization, and urban policy. Antipode, 47(5), 1203–1223.

Dei, G. J. S. (2005). Critical issues in anti-racist research methodology: An introduction. In G. J. S. Dei & G. Johal (Eds.), Critical issues in anti-racist research methodology (pp. 1–28). Peter Lang.

Dei, G. J. S. (2015, January 27). Principles of anti-racist research [Lecture]. Advanced Seminar on Race and Anti-Racism Research Methodology in Education, OISE/ University of Toronto.

Della, T., Adjei, M., Zimmerer, J., & Kamaţuka, N. (2018, April 6). Quo vadis, Hamburg? Zum Stand städtischer, postkolonialer Erinnerungspolitik. Quo Vadis, Hamburg? Der Genozid an den Ovaherero und Nama und die Hamburger Kolonialgeschichte. 2. Transnationaler Herero und Nama Kongress, MS Stubnitz, Hamburg. https://www.youtube.com/watch?v=_HzJGurdukg

Destatis. (2018). Staat & Gesellschaft – Migration & Integration – Personen mit Migrationshintergrund. Destatis: Statistisches Bundesamt. https://www.destatis.de/DE/ZahlenFakten/GesellschaftStaat/Bevoelkerung/MigrationIntegration/Methoden/PersonenMitMigrationshintergrund.html

Deutsche Einheit Fernstraßenplanungs- und -bau GmbH. (n.d.). A 26: Grundsteinlegung für Wasch- und Gebetshaus in Hamburg. DEGES Deutsche Einheit Fernstraßenplanungs- und -bau GmbH. Retrieved June 23, 2019, from https://www.deges.de/Startseite/A-26-Grundsteinlegung-fuer-Wasch-und-Gebetshaus-in-Hamburg-E1702.htm

Deutsche Welle. (2019, June 28). Journalist Deniz Yucel's detention was illegal: Turkish court. Deutsche Welle. https://www.dw.com/en/journalist-deniz-yucels-detention-was-illegal-turkish-court/a-49406261

Deutsche Welle. (2020a, April 21). Germany: Halle synagogue shooter charged with double murder. DW.COM. https://www.dw.com/en/germany-halle-synagogue-shooter-charged-with-double-murder/a-53201696

Deutsche Welle. (2020b, June 28). Germany: Gütersloh slaughterhouse infections spread to community. DW.COM. https://www.dw.com/en/germany-g%C3%BCtersloh-slaughterhouse-infections-spread-to-community/a-53971931

Die Insel Hilft e.V. (n.d.). Die Insel Hilft e.V. - Geflüchteten- und Nachbarschaftshilfe in Hamburg-Wilhelmsburg. Die Insel Hilft e.V. Retrieved June 13, 2019, from https://die-insel-hilft.de/

Dietz, A. (2008). Fremdarbeiter, Gastarbeiter, Einwanderer – Migration in Geschichte und Gegenwart. In Geschichtswerkstatt Wilhelmsburg Honigfabrik e.V. & Museum Elbinsel Wilhelmsburg e. V. (Eds.), Wilhelmsburg – Hamburgs große Elbinsel (pp. 97–112). Medien-Verlag Schubert.

Dörfler, T. (2014, April 9). The governmentalization of gentrification: A case study from Hamburg-Wilhelmsburg on the depriving aspects of diversity management. Governing Urban Diversity: New Challenges for Urban Policy. Association of American Geographers Annual Meeting, Tampa, FL.

Dorries, H., Hugill, D., & Tomiak, J. (2022). Racial capitalism and the production of settler colonial cities. Geoforum, 132, 263-270. https://doi.org/10.1016/j.geoforum.2019.07.016

Drever, A. (2004). Separate spaces, separate outcomes? Neighbourhood impacts on minorities in Germany. Urban Studies, 41(8), 1423–1439.

Eckardt, F. (2017). Architecture and the "Right to the City:" The IBA Hamburg as a case for critical urban studies. In M. Mendes, T. Sá, & J. Cabral (Eds.), Architecture and the social sciences: Inter- and multidisciplinary approaches between society and space (pp. 171–190). Springer.

Ehrkamp, P. (2005). Placing identities: Transnational practices and local attachments of Turkish immigrants in Germany. Journal of Ethnic and Migration Studies, 31(2), 345–364.

Ehrkamp, P. (2006). "We Turks are no Germans": Assimilation discourses and the dialectical construction of identities in Germany. Environment and Planning A, 38, 1673–1692.

Eicke-Diekmann, C., & Popien, M. (2018, November 28). 200 Euro Miete für den Schlafplatz in einer Lagerhalle. Hamburger Abendblatt. https://www.abendblatt.de/hamburg/article215890353/Razzien-gegen-Hauseigentuemer-wegen-illegaler-Vermietungen.html

Eigler, F. (2014). Introduction: Geocritical approaches to place-bound belonging. In Heimat, space, narrative: Toward a transnational approach to flight and expulsion (pp. 1–9). Camden House.

Eigler, F., & Kugele, J. (2012). Introduction: Heimat at the intersection of memory and space (F. Eigler & J. Kugele, Eds.; pp. 1–12). De Gruyter.

Elliott-Cooper, A., Hubbard, P., & Lees, L. (2020). Moving beyond Marcuse: Gentrification, displacement and the violence of un-homing. Progress in Human Geography, 44(3), 492–509.

El-Tayeb, F. (2001). Schwarze Deutsche: Der Diskurs um "Rasse" und nationale Identität 1890-1933. Campus.

El-Tayeb, F. (2005). Dangerous liaisons: Race, nation and German identity. In P. M. Mazón & R. Steingröver (Eds.), Not So Plain as Black and White: Afro-German Culture and History, 1890-2000 (pp. 27–60). Boydell & Brewer.

El-Tayeb, F. (2011). European others: Queering ethnicity in postnational Europe. University of Minnesota Press.

El-Tayeb, F. (2012). 'Gays who cannot properly be gay': Queer Muslims in the neoliberal European city. European Journal of Women's Studies, 19(1), 79–95.

El-Tayeb, F. (2016). Undeutsch: Die Konstruktion des Anderen in der postmigrantischen Gesellschaft. transcript Verlag.

England, M. R., & Simon, S. (2010). Scary cities: Urban geographies of fear, difference and belonging. Social & Cultural Geography, 11(3), 201–207.

European Commission. (2010). Press release – Equality: European Commission closes cases against Germany on anti-discrimination rules. https://europa.eu/rapid/press-release_IP-10-1429_en.htm

Eydlin, A., & AFP. (2020, June 27). Hengameh Yaghoobifarah: Horst Seehofer wertet "taz"-Auseinandersetzung als Erfolg. Die Zeit. https://www.zeit.de/gesellschaft/zeitgeschehen/2020-06/hengameh-yaghoobifarah-horst-seehofer-taz-kolumne-anzeigedrohung-ruecknahme-erfolg

Faas, T., & Wüst, A. M. (2002). The Schill factor in the Hamburg state election 2001. German Politics, 11(2), 1–20.

Federal Anti-Discrimination Agency. (2010). Guide to the General Equal Treatment Act: Explanations and Examples. Federal Anti-Discrimination Agency; https://www.antidiskriminierungsstelle.de/SharedDocs/Downloads/EN/publikationen/agg_wegweiser_engl_guide_to_the_general_equal_treatment_act.pdf?__blob=publicationFile

Forschungsstelle Hamburgs (post-) koloniales Erbe. (2018). "Confronting the Colonial Past!": Impressions from our Conference (28/2/2018-2/3/2018). University of Hamburg.

Forschungsstelle Hamburgs (post-) koloniales Erbe/ Hamburg und die frühe Globalisierung. (n.d.). Hamburgs (post-) koloniales Erbe/ Hamburg und die frühe Globalisierung. Hamburgs (post-)koloniales Erbe. Retrieved November 23, 2018, from https://www.kolonialismus.uni-hamburg.de/

Foucault, M. (1991). Governmentality. In G. Burchell, C. Gordon, & p. Miller (Eds.), The Foucault effect: Studies in governmentality (pp. 87–104). University of Chicago Press.

Foucault, M. (2003). What is critique? In p. Rabinow & N. Rose (Eds.), The essential Foucault: Selections from the essential works of Foucault 1954-1984 (pp. 263–278). The New Press.

Frank, T. (1997). Das Arbeitserziehungslager Wilhelmsburg: Ein Beitrag zum nationalsozialistischen Lagersystem. Geschichtswissenschaft der Universität Hamburg.

Freeman, L. (2006). There goes the 'hood: Views of gentrification from the ground up. Temple University Press.

Freie und Hansestadt Hamburg. (n.d.-a). Das Rahmenprogramm Integrierte Stadtteilentwicklung. Hamburg.de. Retrieved April 19, 2016, from https://www.hamburg.de/rise/

Freie und Hansestadt Hamburg. (n.d.-b). Speicherstadt UNESCO Site Hamburg. Hamburg.de. Retrieved December 30, 2018, from https://www.hamburg.com/sights/speicherstadt/

Freie und Hansestadt Hamburg, Bezirk Hamburg-Mitte. (n.d.). Beteiligung: Bürgerbeteiligung an der Bauleitplanung. Hamburg.de. Retrieved March 28, 2019, from https://www.hamburg.de/mitte/beteiligung/

Freie und Hansestadt Hamburg, Landesbetrieb Geoinformation und Vermessung (2002). Geo-Online. https://geoportal-hamburg.de/geo-online/

Freie und Hansestadt Hamburg, Senatskanzlei. (n.d.). Wohnungsbau: Bestehende politische Instrumente. hamburg.de. Retrieved March 28, 2019, from https://www.hamburg.de/instrumente-wohnungsbau/

Freie und Hansestadt Hamburg, Staatliche Pressestelle. (2002). Leitbild: Metropole Hamburg/ Wachsende Stadt.

Friedrich, L. (2008). Wohnen und innerstädtische Segregation von Migranten in Deutschland (Working Paper 21; Integrationsreport). Bundesamt für Migration und Flüchtlinge.

Fuhlrott, M. (2019, January 30). Über das Kopftuchverbot wird der EuGH entscheiden. Legal Tribune Online. https://www.lto.de/recht/hintergruende/h/kopftuchverbot-bag-europarecht-eugh-religionsfreiheit-freiheit-unternehmer/

Gandy, M. (2002). Concrete and clay: Reworking nature in New York City. MIT Press.

Gaßdorf, U. (2018, April 27). Sozialwohnungen: So können 70.000 Haushalte profitieren. Hamburger Abendblatt. https://www.abendblatt.de/hamburg/kommunales/article214140599/Sozialwohnungen-So-koennen-70-000-Haushalte-profitieren.html

Geschichtswerkstatt Wilhelmsburg und Hafen. (n.d.-a). Entstehung der Elbinsel. Geschichtswerkstatt Wilhelmsburg & Hafen. Retrieved March 9, 2019, from https://www.geschichtswerkstatt-wilhelmsburg.de/entstehung-der-elbinsel/

Geschichtswerkstatt Wilhelmsburg und Hafen. (n.d.-b). Industrialisierung und Stadtentwicklung Wilhelmsburgs im 19. Jahrhundert. Geschichtswerkstatt Wilhelmsburg & Hafen. Retrieved December 12, 2018, from https://www.geschichtswerkstatt-wilhelmsburg.de/geschichte-wilhelmsburgs/

Geschichtswerkstatt Wilhelmsburg und Hafen. (2008). Einwanderer – Einwohner – Einheimische? Begleittext zur gleichnamigen Ausstellung. Geschichtswerkstatt Wilhelmsburg & Hafen. https://www.geschichtswerkstatt-wilhelmsburg.de/geschichte-wilhelmsburgs/

Geschichtswerkstatt Wilhelmsburg und Hafen. (2018). Wo bin ich zu Hause? Die Geschichte der Elbinsel Wilhelmsburg – Band 2. Geschichtswerkstatt Wilhelmsburg und Hafen.

Gilmore, R. W. (2002). Race and globalization. In R. J. Johnston, p. J. Taylor, & M. Watts (Eds.), Geographies of global change: Remapping the world (pp. 261–274). Wiley-Blackwell.

Gilroy, p. (2004). After empire: Melancholia or convivial culture? Routledge.

Gipp, J. (2001, May 23). Suche nach Wegen aus der Misere. Harburger Rundschau.

Goldberg, D. T. (2008). The threat of race: Reflections on racial neoliberalism. Wiley-Blackwell.

González, J. A. (2008). Subject to display: Reframing race in contemporary installation art. The MIT Press.

Gosine, A., & Teelucksingh, C. (2008). Environmental justice and racism in Canada: An introduction. Emond Montgomery Publications.

Gözübüyük, D., & Weinhold, M. (2017, July 22). Diskriminierung im Alltag: Warum Max eine Wohnung bekommt – und Daniel nicht. MOPO.de. https://www.mopo.de/hamburg/diskriminierung-im-alltag-warum-max-eine-wohnung-bekommt---und-daniel-nicht-28006054

Greven, L. (2014, April 5). Wilhelmsburg: Hamburgs wildes Stadtlabor. Die Zeit. https://www.zeit.de/hamburg/stadtleben/2014-04/hamburg-wilhelmsburg

Greven, L. (2015, January 13). Gentrifizierung: Wilhelmsburg will nicht hip werden. Die Zeit. https://www.zeit.de/hamburg/stadtleben/2015-01/hamburg-wilhelmsburg-gentrifizierung-arm

Gruner, S. (2010). 'The others don't want ...'. Small-scale segregation: Hegemonic public discourses and racial boundaries in German neighbourhoods. Journal of Ethnic and Migration Studies, 36(2), 275–292.

Güntner, S. (2013). Kalkulierbare Segregation? Drei Perspektiven auf die sozialräumliche Polarisierung in Hamburg. In G. Pohl & K. Wicher (Eds.), Hamburg: Gespaltene Stadt? Soziale Entwicklungen in der Metropole (pp. 34–53). VSA: Verlag.

Güvercin, E. (2010). Ein Teil Deutschlands mit etwas mehr Farbe. Aus Politik und Zeitgeschichte: Anerkennung, Teilhabe, Integration, 46–47, 3–9.

Ha, K. N. (2003). Die kolonialen Muster deutscher Arbeitsmigrationspolitik. In H. Steyerl & E. Gutiérrez Rodríguez (Eds.), Spricht die Subalterne Deutsch? Migration und postkoloniale Kritik (pp. 56–107). Unrast-Verl.

Ha, K. N. (2005). Macht(t)raum(a) Berlin: Deutschland als Kolonialgesellschaft. In M. M. Eggers, G. Kilomba, p. Piesche, & S. Arndt (Eds.), Mythen, Masken und Subjekte: Kritische Weißseinsforschung in Deutschland (pp. 105–117). Unrast-Verl.

Ha, K. N. (2007a). Deutsche Integrationspolitik als koloniale Praxis. In N. Lauré al-Samarai, S. Mysorekar, & K. N. Ha (Eds.), re/visionen: Postkoloniale perspektiven von People of Color auf Rassismus, Kulturpolitk and Widerstand in Deutschland (pp. 113–128). Unrast-Verl.

Ha, K. N. (2007b). Koloniale Arbeitsmigrationspolitik im Imperial Germany. In K. N. Ha, N. Lauré al-Samarai, & S. Mysorekar (Eds.), Re/visionen: Postkoloniale perspektiven von People of Color auf Rassismus, Kulturpolitk and Widerstand in Deutschland (pp. 65–71). Unrast-Verl.

Ha, N. (2014a). Perspektiven urbaner Dekolonisierung: Die europäische Stadt als ‚Contact Zone'. sub\urban. zeitschrift für kritische stadtforschung, 2(1), 27–48.

Ha, N. (2014b). Zur Aktualität von städtischem Orientalismus: Zur Zulässigkeit von Diskriminierung auf dem Wohnungsmarkt. Leben Nach Migration: Newsletter Des Migrationsrates Berlin Brandenburg, 3, 17–18.

Ha, N. (2016). Straßenhandel in Berlin: Öffentlicher Raum, Informalität und Rassismus in der neoliberalen Stadt. transcript Verlag.

Ha, N. (2017). Zur Kolonialität des Städtischen. In Zwischenraum Kollektiv (Ed.), Decolonize the City! (pp. 75–87). Unrast-Verl.

Ha, N., & Picker, G. (Eds.). (2022). European cities: Modernity, race and colonialism. Manchester University Press.

Hackworth, J. (2002). Postrecession gentrification in New York City. Urban Affairs Review, 37(6), 815–843.

Haddad, L. (2018). Anerkennung und Widerstand: Lokale islamische Identitätspraxis in Hamburg. transcript Verlag.

HafenCity Hamburg GmbH. (n.d.-a). HafenCity Hamburg – Home. Retrieved January 24, 2019, from https://www.hafencity.com/en/home.html

HafenCity Hamburg GmbH. (n.d.-b). HafenCity Hamburg – On historic ground. Retrieved March 5, 2019, from https://www.hafencity.com/en/overview/on-historic-ground.html

Hall, S. (1986). Gramsci's relevance for the study of race and ethnicity. Journal of Communication Inquiry, 10(2), 5–27.

Hall, S. (1997). The spectacle of the "other." In S. Hall (Ed.), Representation: Cultural representations and signifying practices (pp. 223–290). Sage Publications.

Hall, S. (2002). "In but not of Europe:" Europe and its myths. Soundings: A Journal of Politics and Culture, 22(Winter), 57–69.

Hall, S., Critcher, C., Jefferson, T., Clarke, J., & Roberts, B. (1978). Policing the crisis: Mugging, the state and law and order. Palgrave/MacMillan.

Hamburg für gute Integration. (n.d.). Volksinitiative. Hamburg für gute Integration. Retrieved June 13, 2019, from https://www.gute-integration.de/

Hamburger Abendblatt. (2000, September 22). "Hilferuf aus der Bronx" verärgert Ortsausschuss. https://www.abendblatt.de/archiv/2000/article204341787/Hilferuf-aus-der-Bronx-veraergert-Ortsausschuss.html

Hamburger Abendblatt. (2010, February 18). 3300 Bäume sollen weichen – Protest gegen "Kahlschlag." https://www.abendblatt.de/hamburg/kommunales/article107647965/3300-Baeume-sollen-weichen-Protest-gegen-Kahlschlag.html

Hamburger Abendblatt. (2016, May 29). Grüne und Muslime wollen Stadtteilmoscheen in Hamburg. Hamburger Abendblatt. https://www.abendblatt.de/hamburg/article207616461/Gruene-und-Muslime-wollen-Stadtteilmoscheen-in-Hamburg.html

Hamburger Abendblatt. (2017, December 20). Nach Explosion: Mann mit rechter Vergangenheit in Haft. Hamburger Abendblatt. https://www.abendblatt.de/hamburg/article212888499/Explosion-am-S-Bahnhof-Veddel-Tatverdaechtiger-festgenommen.html

Hamburgische Investitions- und Förderbank. (2016). Wohnen für Studierende und Auszubildende in Hamburg-Süd: Förderrichtlinie zur Förderung des Wohnens für Studierende und Auszubildende in einzelnen Quartieren im Süden Hamburgs.

Hammersley, M., & Atkinson, P. (2007). Ethnography: Principles in practice. Routledge.

Handelskammer Hamburg. (n.d.). Hamburg: A key financial centre. Handelskammer Hamburg. Retrieved December 7, 2018, from https://www.hk24.de/en/produktmarken/financial-economy/hamburg-key-financial-centre/1147600

Haritaworn, J. (2005). Am Anfang war Audre Lorde: Weißsein und Machtvermeidung in der queeren Ursprungsgeschichte. Femina Politica – Zeitschrift für feministische Politikwissenschaft, 14(1), 23–35.

Haritaworn, J. (2012). Biopolitics of mixing: Thai multiracialities and haunted ascendancies. Ashgate Publishing.

Haritaworn, J. (2015). Queer lovers and hateful others: Regenerating violent times and places. Pluto Press.

Harlander, T. (2012). Zuwanderung und 'überforderte Nachbarschaften' – Zu Debatte um Segregation und Integration seit 1989. In T. Harlander & G. Kuhn (Eds.), Soziale Mischung in der Stadt: Case Studies – Wohnungspolitik in Europa – Historische Analyse (pp. 306–313). Wüstenrot Stiftung / Karl Krämer Verlag.

Harlander, T., & Kuhn, G. (2012). Deutschland: Mischung in kommunaler Praxis und Wohnungswirtschaft. In Soziale Mischung in der Stadt: Case Studies – Wohnungspolitik in Europa – Historische Analyse (pp. 386–407). Wüstenrot Stiftung / Karl Krämer Verlag.

Hartmann, B., Klay, A., & Kranz, B. (2019, January 31). Kopftuch-Verbot: Dürfen Unternehmen die Kleidung vorschreiben? Hamburger Abendblatt. https://www.abendblatt.de/wirtschaft/article216337923/Kopftuch-Verbot-Duerfen-Chefs-ueber-Kleidung-bestimmen.html

Häußermann, H. (2001). Die europäische Stadt. Leviathan, 29(2), 237–255.

Hellweg, U. (2010). Projects for the future of the metropolis: IBA Hamburg 2013. In E. Lütke Daldrup, p. Zlonicky, & Federal Ministry of Transport, Building and Urban Affairs (Eds.), & J. Lutes (Trans.), Large scale projects in German cities: Urban development 1990-2010 (pp. 114–119). Jovis Verlag.

Hellweg, U. (2013). IBA Hamburg: An initial review. In Metropolis: Building the city anew (pp. 20–47). Jovis Verlag and IBA Hamburg GmbH.

Henderson, H. (2016). Toward an ethnographic sensibility in urban research. Australian Planner, 53(1), 28–36.

Hengari, M., Murangi, K., & Mancheno, T. (2018, April 7). Kolonial Amnesie: Hamburg als Profiteur deutscher Kolonialpolitik im heutigen Namibia. Quo Vadis, Hamburg? Der Genozid an den Ovaherero und Nama und die Hamburger Kolonialgeschichte. 2. Transnationaler Herero und Nama Kongress, University of

Hamburg. https://www.youtube.com/watch?v=YW3Mrf17FMQ&list=PLmnAZl zZZovKp-RxC9BVbOG59VXPEscos&index=5

Hildebrand, U. (2003). Heimat ist, wo ich mich wohlfühle: Themenblätter im Unterricht. Bundeszentrale für politische Bildung.

Hill, J. (2019, October 11). Germany shooting: Horror of Jews targeted in anti-Semitic attack. BBC News. https://www.bbc.com/news/world-europe-499977 79

Hinze, A. M. (2013). Turkish Berlin: Integration policy and urban space. University of Minnesota Press.

Hohenstatt, F. (2017). Städtische Räume zwischen Aufwertung und Ausschliessung: Zur Position Sozialer Arbeit in einem "Regime der Umstrukturierung" [Doctoral Dissertation]. Johann Wolfgang Goethe Universität zu Frankfurt am Main.

Hohenstatt, F., & Rinn, M. (2013). Festivalisierte Problemarbeitung: Die bevölkerungspolitische Strategie der IBA Hamburg, die Abwesenheit Sozialer Arbeit in Stadtentwicklungspolitik und die Effekte auf Wohnverhältnisse in Wilhelmsburg. Widersprüche, 127(1), 23–38.

Holm, A. (2009). Soziale Mischung. Zur Entstehung und Funktion eines Mythos. Forum Wissenschaft, January 2009. https://www.bdwi.de/forum/archiv/archi v/2380964.html.

hooks, b. (1992). Black looks: Race and representation. South End Press.

Housing Europe. (2015, June 16). "Social mixing is not a one-size-fits-all solution to urban problems." Housing Europe. https://www.housingeurope.eu/resourc e-499/social-mixing-is-not-a-one-size-fits-all-solution-to-urban-problems

Humburg, M. (2009). Wem gehören die Elbinseln? Wilhelmsburger Insel Rundblick, 3–6.

Humburg, M., & Rothschuh, M. (2018). Wenn die Kommune verschwindet… Bürgerengagement und Bürgerbeteiligung in Wilhelmsburg – Einem Hamburger Stadtteil ohne eigene Verwaltung und politische Vertretung. Netzwerk Bürgerbeteiligung, 03–2018.

Hunting, G. (n.d.). Intersectionality-informed qualitative research: A primer. Simon Fraser University.

I. Transnational Congress on the Ovaherero and Nama Genocide. (2016). Berlin Resolution 2016: Restorative Justice after Genocide. Berlin. https://genocide-n amibia.net/2016/10/berlin-resolution-2016/

IBA Hamburg. (n.d.-a). Internationale Gartenschau igs hamburg 2013 GmbH. Retrieved April 5, 2019, from https://www.iba-hamburg.de/story/akteure/igs-ha mburg-2013-gmbh.html

IBA Hamburg. (n.d.-b). Projekte A-Z. IBA Hamburg. Retrieved January 20, 2019, from https://www.iba-hamburg.de/nc/projekte/projekte-a-z.html

IBA Hamburg. (n.d.-c). Spreehafenviertel. IBA Hamburg. Retrieved October 25, 2019, from https://www.iba-hamburg.de/de/projekte/spreehafenviertel/uebersicht

IBA Hamburg. (2010). Projects and concepts: Catalogue for the interim presentation 2010. Jovis Verlag and IBA Hamburg GmbH.

IBA Hamburg. (2011a). Mission: To show the future. IBA Hamburg. https://www.iba-hamburg.org/en/01_entwuerfe/3_mission/mission_zukunftzeigen.php

IBA Hamburg. (2011b). The great flood and its consequences. History of the Elbe islands: Twelve leaps in time. IBA Hamburg. https://www.iba-hamburg.org/en/01_entwuerfe/5_schauplatz/schauplatz_historie_1962.php

IBA Hamburg. (2013). Gentrifizierung in Wilhelmsburg? Internationale Bauausstellung Hamburg.

IBA Hamburg. (2015). Format IBA. https://www.iba-hamburg.de/en/story/format-iba.html

IBA Hamburg. (2016). IBA Hamburg – Projects A-Z. IBA Hamburg. https://www.iba-hamburg.de/en/nc/projects/projects-a-z.html

IBA Hamburg. (2018). IBA Hamburg – Seven years on the island. IBA Hamburg. https://www.iba-hamburg.de/intro.html

IBA Hamburg. (2019). Startseite. IBA Hamburg. IBA Hamburg. https://www.iba-hamburg.de/de/

IBA Hamburg. (2020). Universität der Nachbarschaften. IBA Hamburg. https://www.internationale-bauausstellung-hamburg.de/projekte/universitaet-der-nachbarschaften/projekt/universitaet-der-nachbarschaften.html

Iken, M. (2020, June 18). Heute ein Bismarck-Denkmal aufstellen? Wohl kaum. Hamburger Abendblatt, 14.

Initiativen für erfolgreiche Integration. (n.d.). Forderungen. Initiativen für erfolgreiche Integration Hamburg. Retrieved June 13, 2019, from https://www.ifi-hamburg.de/über-ifi-hamburg/forderungen/

Institut für Migrations- und Rassismusforschung Hamburg. (2011). European Network Against Racism shadow report 2009/2010: Racism and discrimination in Germany. European Network Against Racism.

James, C. (2012). Life at the intersection: Community, class and schooling. Fernwood.

Journalisten der Henri-Nannen-Schule. (2015, May 6). Sinti in Hamburg: Emils Ring. Die Zeit. https://www.zeit.de/zeit-magazin/leben/2014-07/sinti-hamburg-emil-weiss

Kaddor, L. (2010). Muslimisch – Weiblich – Deutsch! Mein Weg zu einem zeitgemäßen Islam. Bundeszentrale für politische Bildung.

Kampagne für die Grundrechte. (2013). Gefahrengebiete in Hamburg. Kampagne Für Die Grundrechte. https://www.grundrechte-kampagne.de/map/node

Karakayali, J., & Nieden, B. zur. (2013). Rassismus und Klassen-Raum: Segregation nach Herkunft an Berliner Grundschulen. sub\urban. zeitschrift für kritische stadtforschung, 1(2), 61–78.

Kaschel, H. (2019, October 5). Deniz Yücel: Erdogan let me be tortured in Turkish prison. Deutsche Welle. https://www.dw.com/en/deniz-y%C3%BCcel-erdogan-let-me-be-tortured-in-turkish-prison/a-48686030

Kauffman, J. (2015). Elusive alliance: The German occupation of Poland in World War I. Harvard University Press.

Kawash, S. (1998). The homeless body. Public Culture, 10(2), 319–339. https://doi.org/10.1215/08992363-10-2-319

Keesenberg, H. (1989). Insel der Gegensätze. Hiltrud Tiedemann.

Keller, C. (2015). Problemviertel? Imageproduktion und soziale Benachteiligung städtischer Quartiere. Bundeszentrale für politische Bildung. https://www.bpb.de/politik/innenpolitik/gangsterlaeufer/202834/problemviertel-image-und-benachteiligung

Kellermann, K. (2019). Silence, motifs and echoes: Acts of listening in postcolonial Hamburg. In P. Hildebrandt, K. Evert, S. Peters, M. Schaub, K. Wildner, & G. Ziemer (Eds.), Performing citizenship: Bodies, agencies, limitations (pp. 93–109). Palgrave Macmillan.

Kheraj, S. (2013). Inventing Stanley Park: An environmental history. UBC Press.

King, A. D. (1990). Urbanism, colonialism, and the world-economy (Reprint edition). Routledge.

Kipfer, S. (2016). Neocolonial urbanism? La rénovation urbaine in Paris. Antipode, 48(3), 603–625.

Kipfer, S., & Goonewardena, K. (2014). Henri Lefebvre and "colonization": From reinterpretation to research. In C. Schmid, L. Stanek, & A. Moravanszky (Eds.), Urban revolution now: Henri Lefebvre in social research and architecture (pp. 93–112). Routledge.

Kirsche, G. (2018, January 4). Nur eine kleine Detonation.jungle.world. https://jungle.world/artikel/2018/01/nur-eine-kleine-detonation

Kleinhubbert, G. (2008, April 3). Clash der Kulturen: Jungakademiker erobern Problemviertel. Spiegel Online. https://www.spiegel.de/lebenundlernen/uni/clash-der-kulturen-jungakademiker-erobern-problemviertel-a-536106.html

Knödler, G. (2021, April 2). Sinti-Siedlung in Hamburg-Wilhelmsburg: Neustart nach 40 Jahren. Die Tageszeitung: taz. https://taz.de/!5758502/

Kopf, B. (2012). Warum ich gern in Wilhelmsburg lebe – Und nicht ganz woanders. In Zukunft Elbinsel Wilhelmsburg (Ed.), Eine Starke Insel mitten in der Stadt (p. 102). Zukunft Elbinsel Wilhelmsburg e. V.

Kopp, K. (2010). Gray zones: On the inclusion of "Poland" in the study of German colonialism. In M. Perraudin & J. Zimmerer (Eds.), German colonialism and national identity (pp. 33–41). Taylor & Francis.

Krieger, S., Schöneich, C., Kortmann, K., & Kott, K. (2021). Wohnverhältnisse. Bundeszentrale für politische Bildung. https://www.bpb.de/kurz-knapp/zahlen-und-fakten/datenreport-2021/wohnen/329992/wohnverhaeltnisse/

Krohn, p. (2018, July 7). Ein Jahr nach dem G-20-Gipfel: Wann ist die Schanze gekippt? FAZ.NET. https://www.faz.net/aktuell/gesellschaft/menschen/wann-ist-die-schanze-gekippt-15676801.html

Kuppinger, p. (2011). Vibrant mosques: Space, planning and informality in Germany. Built Environment, 37(1), 78–91.

Kuppinger, p. (2014). Mosques and minarets: Conflict, participation, and visibility in German cities. Anthropological Quarterly, 87(3), 793–818.

Landesbetrieb Straßen, Brücken und Gewässer. (n.d.). Hamburgs längste Sitzbank schaut aufs Wasser: Die ersten 500 Meter des Klütjenfelder Hauptdeich wurden um 80 Zentimeter erhöht. hamburg.de. Retrieved November 16, 2019, from https://lsbg.hamburg.de/np-service/13203524/erhoehung-kluetjenfelder-hauptdeich/

Landeszentrale für Politische Bildung Hamburg. (2007). Zwangsarbeit in der Hamburger Kriegswirtschaft 1939-1945. https://www.zwangsarbeit-in-hamburg.de

Lanz, S. (2002). Mythos europäische Stadt – Fallstricke aktueller Rettungsversuche. In WD. Bukow & E. Yildiz (Eds.), Der Umgang mit der Stadtgesellschaft (Vol. 11, pp. 63–77). VS Verlag für Sozialwissenschaften.

Lasarzik, A. (2014, May 14). Gefahrengebiet Inselpark. Mittendrin: Das Nachrichtenmagazin für Hamburg-Mitte. https://hh-mittendrin.de/2014/05/gefahrengebiet-inselpark/

Lauterbach, J. (2011, August 29). Hamburger Terrorzelle: Nachmieter in der Wohnung von Mohammed Atta. Die Welt. https://www.welt.de/politik/deutschland/article13572179/Nachmieter-in-der-Wohnung-von-Mohammed-Atta.html

Lees, L. (2008). Gentrification and social mixing: Towards an inclusive urban renaissance? Urban Studies, 45(12), 2449–2470.

Leoni, A. (2018, December 6). Hamburger Kita kündigt Erzieherin, weil sie Kopftuch trägt. Hamburger Abendblatt. https://www.abendblatt.de/hamburg/article215954427/Hamburger-Kita-kuendigt-Erzieherin-weil-sie-Kopftuch-traegt.html

Lewicka, M. (2011). Place attachment: How far have we come in the last 40 years? Journal of Environmental Psychology, 31(3), 207–230.

Lichtblau, Q. (2019, March 21). „Von uns aus können wir den Heimatbegriff den Rechten überlassen: "Interview mit Fatma Aydemir und Hengameh Yaghoobifarah zu ihrem Buch "Eure Heimat ist unser Albtraum." Jetzt. https://www.jetzt.de/politik/interview-mit-fatma-aydemir-und-hengameh-yaghoobifarah-zu-ihrem-buch-eure-heimat-ist-unser-albtraum

Linksunten HH. (2017, May 2). Spontandemonstration und Rache Aktion am 1. Mai in Wilhelmsburg in Solidarität mit dem Kurdischen Widerstandskampf!!! Linksunten. https://linksunten.indymedia.org/en/node/211180

Lintschnig, H. (2018, August). „Jetzt sind mehr Deutsche hier“: Julie Chamberlain aus Kanada schreibt eine Doktorarbeit über Veränderungen in Wilhelmsburg seit der IBA - aus Sicht von Migrant*innen. Wilhelmsburger Insel Rundblick, 12–13.

Lintschnig, H., & Schultz, S. (2018, October 21). Bulgarische Billiglöhner in Wilhelmsburg: Auf dem Arbeiterstrich. Spiegel Online. https://www.spiegel.de/wirtschaft/soziales/arbeitsmigration-wie-bulgaren-in-hamburg-ausgebeutet-werden-a-1232762.html

Lowe, L. (2015). The Intimacies of Four Continents. Duke University Press.

Lütke Daldrup, E., & Zlonicky, P. (2010). Large-scale projects in German cities: Urban development 1990-2010. (J. Lutes, Trans.). Jovis Verlag.

Lutolli, B. (2021). An evaluation of housing flexibility after seven years of habitation: IBA Hamburg 2013. *Journal of Architecture and Urbanism*, 45(2), 195–204. https://dx.doi.org/10.3846/jau.2021.14515

MacDougall, C. E. (2011). Cold War capital: Contested urbanity in West Berlin, 1963–1989 [Doctoral Dissertation]. Rutgers, The State University of New Jersey.

MacFarlane, K. (2019, July 18). Marshland: Hamburg G20 and the return of the Hanseatic League. Society & Space. https://societyandspace.org/2019/07/18/marshland-hamburg-g20-and-the-return-of-the-hanseatic-league/

MacFarlane, K., & Mitchell, K. (2019). Hamburg's spaces of danger: Race, violence and memory in a contemporary global city. International Journal of Urban and Regional Research, 43(5), 816–832.

Maeck, S. (2009, August 7). Elisabeth und ihr goldenes Kalb in der Elbe. https://www.abendblatt.de/region/stade/article107542282/Elisabeth-und-ihr-goldenes-Kalb-in-der-Elbe.html

Mancheno, T. (2016, June 29). Das postkoloniale Erbe der Hafencity. Hamburg: Deutschlands Tor zur kolonialen Welt. Über den Umgang mit einem schwierigen Erbe., University of Hamburg. https://www.kolonialismus.uni-hamburg.de/2016/07/04/video-tania-mancheno-universitaet-hamburg-das-postkoloniale-erbe-der-hafencity/

Marinov, V. (2016, October 6). Flüchtlinge ziehen in die HafenCity: Dieses Containerdorf hat den schönsten Elphi-Blick. MOPO.de. https://www.mopo.de/hamburg/fluechtlinge-ziehen-in-die-hafencity-dieses-containerdorf-hat-den-schoensten-elphi-blick-24853118

May, R. A. B. (2014). Urban nightlife: Entertaining race, class, and culture in public space. Rutgers University Press.

McCreary, T., & Milligan, R. (2021). The limits of liberal recognition: Racial capitalism, settler colonialism, and environmental governance in Vancouver and Atlanta. Antipode, 53 (3), 724-744.

McDermott, T. (2002, January 27). A Perfect Soldier. Los Angeles Times. https://articles.latimes.com/2002/jan/27/news/mn-25005

Melamed, J. (2015). Racial capitalism. Critical Ethnic Studies, 1(1), 76–85.

Mele, C. (2015). Contemporary urban development and compounded exclusion. In R. M. Adelman & C. Mele (Eds.), Race, space, and exclusion: Segregation and beyond in metropolitan America (pp. 75–89). Routledge.

Menzel, J. (2010). Wie nachhaltig ist die HafenCity Hamburg? Zukunftsrat Hamburg.

Meyer-Lenz, J. (2016). Hafenentwicklung: Entwicklung des Hamburger Hafens zum modernen Überseehafen. Geschichtsbuch Hamburg. https://geschichtsbuch.hamburg.de/epochen/kaiserreich/hafenentwicklung/

Meyer-Lenz, J. (2018). Einleitung: Migration, Transnationalismus und Gender: Ein Beitrag zur Forschungsdiskussion. In J. Meyer-Lenz & N. M. Fahnenbruck (Eds.), Fluchtpunkt Hamburg: Zur Geschichte von Flucht und Migration in Hamburg von der frühen Neuzeit bis zur Gegenwart. transcript Verlag.

Meyer-Wellmann, J. (2000, September 5). Wilhelmsburg ruft um Hilfe. Hamburger Abendblatt. https://www.abendblatt.de/archiv/2000/article204328649/Wilhelmsburg-ruft-um-Hilfe.html

Mirchandani, K., Ng, R., Coloma-Moya, N., Maitra, S., Rawlings, T., Shan, H., Siddiqui, K., & Slade, B. (2011). The entrenchment of racial categories in precarious employment. In D. Glenday & N. Pupo (Eds.), The shifting landscape of work (pp. 119–138). Nelson Educational Ltd.

Mobiles Beratungsteams gegen Rechstextremismus. (2018). Monitoring (Extrem) rechte Strukturen und das zivilgesellschaftliche Gegenengagement Nr.9 (Nr. 9). MBT Mobiles Beratungsteams gegen Rechstextremismus.

Mohanram, R. (1999). Black body: Women, colonialism, and space. University of Minnesota Press.

Münch, S. (2009). "It's all in the mix": Constructing ethnic segregation as a social problem in Germany. Journal of Housing and the Built Environment, 24(4), 441.

Münch, S. (2010). Integration durch Wohnungspolitik? Zum Umgang mit ethnischer Segregation im europäischen Vergleich. VS Verlag für Sozialwissenschaften;

Münch, S. (2014). Das „Mantra der Mischung": Die Problematisierung von ethnischer Segregation in Deutschland und den Niederlanden. In p. Gans (Ed.), Räumliche Auswirkungen der internationalen Migration (pp. 327–343). Akademie für Raumforschung und Landesplanung.

Nack, C. (2020, June 28). Europäische Fleischindustrie als Corona-Infektionsherd. DW.COM. https://www.dw.com/de/europ%C3%A4ische-fleischindustrie-als-corona-infektionsherd/a-53955048

Nast, J., & Blokland, T. (2014). Social mix revisited: Neighbourhood institutions as setting for boundary work and social capital. Sociology, 48(3), 482–499.

Neuber, H. (2020, June 27). Umstrittener Polizeieinsatz in Hamburg: Jetzt sprechen die Betroffenen. Telepolis. https://www.heise.de/tp/features/Umstrittener-Polizeieinsatz-in-Hamburg-Jetzt-sprechen-die-Betroffenen-4797186.html

Neue Deutsche Organisationen. (n.d.). Retrieved April 18, 2018, from https://neue-deutsche-organisationen.de/de/

neue deutsche organisationen. (2017). Gleich ≠ Gleich: Antidiskriminierungs und Gleichstellungsdaten im Gespräch. neue deutsche organisationen.

Neues Deutschland. (2014, April 3). Arme, Alte und Ausländer werden verdrängt. Neues Deutschland. https://www.neues-deutschland.de/artikel/925833.arme-alte-und-auslaender-werden-verdraengt.html

Niewyk, D. L., & Nicosia, F. R. (2012). The Columbia guide to the Holocaust. Columbia University Press.

Nightingale, C. H. (2012). Segregation: A global history of divided cities. University of Chicago Press.

Norddeutscher Rundfunk. (2018, April 2). Hamburg arbeitet koloniales Erbe auf. Norddeutscher Rundfunk: Hamburg Journal.

OECD. (2018). PISA: Equity in Education: Breaking down barriers to social mobility: Germany [Country Note].

Open Society Institute. (2010). Muslime in Hamburg. Open Society Institute.

OvaHerero, Mbanderu and Nama Genocides Institute. (2018, May 9). Letter to Dr. Carsten Brosda, Senator of the Free and Hanseatic City of Hamburg. https://theongi.org/?p=1194

Paech, F. (2008). "Die ganzen menschlichen Geschichten:" Die Hamburger Sturmflut von 1962 im Bewusstsein der Wilhelmsburg Bevölkerung. In Geschichtswerkstatt Wilhelmsburg Honigfabrik e.V. & Museum Elbinsel Wilhelmsburg e. V. (Eds.), Wilhelmsburg – Hamburgs große Elbinsel (pp. 161–174). Medien-Verlag Schubert.

Pain, R. (2000). Place, social relations and the fear of crime: A review. Progress in Human Geography, 24(3), 365–387.

Pain, R. (2001). Gender, race, age and fear in the city. Urban Studies, 38(5–6), 899–913.

Patterson, T. (2016, April 8). Rich people in Germany are using limousines and yachts to block a refugee hostel being built. The Independent. https://www.independent.co.uk/news/world/europe/refugee-crisis-hamburg-s-wealthy-take-to-the-streets-in-posh-protest-at-planned-migrant-hostel-a6973771.html

Petzen, J. L. (2009). Gender politics in the new Europe: "Civilizing" Muslim sexualities [Doctoral Dissertation]. University of Michigan.

Pink, S. (2007). Doing visual ethnography. Sage Publications.

Pink, S. (2008). Mobilising visual ethnography: Making routes, making place and making images. Forum Qualitative Sozialforschung/ Forum: Qualitative Social Research, 9(3).

Pink, S. (2009). Doing sensory ethnography. Sage Publications.

Pohlan, J. (2018, May 16). Hamburg unter Beobachtung: Das Sozialmonitoring im Rahmenprogramm Integrierte Stadtteilentwicklung. CCG Ringvorlesung Stadt und Gesundheit, HAW Hamburg.

Popp, M., & Gezer, Ö. (2013, May 9). The usual suspects: Hamburg police accused of racial profiling. Der Spiegel. https://www.spiegel.de/international/germany/hamburg-neighborhood-altona-complains-of-racial-profiling-by-police-a-920440.html

Preißler, S. (2019, February 28). Rekord: Einfamilienhäuser in Hamburg für 1,7 Milliarden verkauft. Hamburger Abendblatt. https://www.abendblatt.de/wirtschaft/article216546193/Rekord-Einfamilienhaeuser-fuer-1-7-Milliarden-Euro-verkauft.html

Pro Asyl. (2017, December 28). Gewalt gegen Flüchtlinge 2017: Von Entwarnung kann keine Rede sein. https://www.proasyl.de/news/gewalt-gegen-fluechtlinge-2017-von-entwarnung-kann-keine-rede-sein/

Prozessbeobachtung zum rechten Terroranschlag auf der Veddel. (2018, December 22). 19. Verhandlungstag – „Zehn Jahre wegen versuchtem Mordes, ein 'ausländerfeindliches' Motiv liegt nahe". Prozessbeobachtung zum rechten Terroranschlag auf der Veddel. https://veddelprozess.blogsport.eu/category/prozessberichte/

Pulido, L. (2006). Black, brown, yellow, and left: Radical activism in Los Angeles. University of California Press.

Pulido, L. (2008). FAQs: Frequently (un)asked questions about being a scholar activist. In C. R. Hale (Ed.), Engaging Contradictions: Theory, Politics, and Methods of Activist Scholarship (pp. 341–365). GAIA Books.

Pulido, L. (2015). Geographies of race and ethnicity 1: White supremacy vs white privilege in environmental racism research. Progress in Human Geography, 39(6), 809–817.

Pulido, L. (2016). Flint, environmental racism, and racial capitalism. Capitalism Nature Socialism, 27(3), 1–16.

Pulido, L. (2017). Geographies of race and ethnicity II: Environmental racism, racial capitalism and state-sanctioned violence. Progress in Human Geography, 41(4), 524–533.

Raddatz, L., & Mennis, J. (2013). Environmental justice in Hamburg, Germany. The Professional Geographer, 65(3), 495–511.

Ramm, C. (2010). The Muslim makers. Interventions, 12(2), 183–197.

Ranganathan, M. (2016). Thinking with Flint: Racial liberalism and the roots of an American water tragedy. Capitalism Nature Socialism, 27(3), 17–33.

Räthzel, N. (1994). Harmonious 'Heimat' and Disturbing 'Auslander.' Feminism & Psychology, 4(1), 81–98.

Razack, S. (2002). When place becomes race. In S. Razack (Ed.), Race, space and the law: Unmapping a white settler society (pp. 1–20). Between the Lines.

Razack, S. (2008). Casting out: The eviction of Muslims from western law and politics. University of Toronto Press.

Rebaschus, M. (2005, December 30). Wilhelmsburg soll wie Kreuzberg werden. Hamburger Abendblatt. https://www.abendblatt.de/hamburg/article107072564/Wilhelmsburg-soll-wie-Kreuzberg-werden.html

Recht auf Stadt – Netzwerk Hamburg. (n.d.). Recht auf Stadt – Netzwerk Hamburg. Retrieved April 11, 2019, from https://www.rechtaufstadt.net/

Riebe, S., & Röer, M. (2019, September 9). Baum-Besetzung in Hamburg: Update: Zu gefährlich! Polizei bricht Rettungsaktion ab. MOPO.de. https://www.mopo.de/hamburg/polizei/baum-besetzung-in-hamburg-update--zu-gefaehrlich--polizei-bricht-rettungsaktion-ab-33135790

Rinn, M. (2018). Ein Urbanismus der Ungleichheit: ‚Neue soziale Stadtpolitik' in Hamburg als Strategie der Verbürgerlichung. Sub\urban. Zeitschrift Für Kritische Stadtforschung, 6(1), 9–28.

Robinson, C. (1983). Black Marxism: The making of the Black radical tradition. University of North Carolina Press.

Rodatz, M. (2012). Produktive "Parallelgesellschaften". Migration und Ordnung in der (neoliberalen) "Stadt der Vielfalt." BEHEMOTH - A Journal on Civilisation, 2(1), 70–103.

Romani Kafava Wilhelmsburg. (n.d.). Romani Kafava Wilhelmsburg. Facebook. https://www.facebook.com/pages/category/Community/Romani-Kafava-Wilhelmsburg-782494631954604/

Römhild, R. (2018). Global Heimat: (Post)migrant productions of transnational space. Anthropological Journal of European Cultures, 27(1), 27–39. https://doi.org/10.3167/ajec.2018.270106

Rowe, M. (2015, February). Building a future. Geographical, 38–43. https://geographical.co.uk/magazine/issues/item/690-february-2015.

Roy, A. (2018, February 13). At the limits of urban theory: Racial banishment in the contemporary city. https://www.lse.ac.uk/lse-player/home.aspx

Ruiz-Tagle, J. (2016). The broken promises of social mix: The case of the Cabrini Green/Near North area in Chicago. Urban Geography, 37(3), 352–372.

Sachs, M. (2012). Interview. In T. Harlander & G. Kuhn (Eds.), Soziale Mischung in der Stadt: Case Studies – Wohnungspolitik in Europa – Historische Analyse (pp. 409–410). Wüstenrot Stiftung / Karl Krämer Verlag.

Said, E. (1979). Orientalism. Vintage Books.

Sanyal, M. (2019a). Zuhause. In F. Aydemir & H. Yaghoobifarah (Eds.), Eure Heimat ist unser Albtraum (pp. 101–121). Ullstein Buchverlage.

Sanyal, M. (2019b, September 18). Suddenly, it's OK to be German and to talk about race. The Guardian. https://www.theguardian.com/commentisfree/2019/sep/18/germany-race-conversation-afd-openness

Sanyal, M. (2020, March 3). The Hanau terror attack shows the need for honesty about racism in Germany. The Guardian. https://www.theguardian.com/world/commentisfree/2020/mar/03/hanau-terror-attack-germany-racism-angela-merkel

Sassen, S. (2013, July 26). To AKU: Response to open letter. https://akuwilhelmsburg.blogsport.eu/2013-07-antwort-von-saskia-sassen-auf-offenen-brief/

Saunders, D. (2010). Arrival city: The final migration and our next world (1st ed.). Alfred A. Knopf Canada.

Saville-Smith, K., Saville-Smith, N., & James, B. (2015). Neighbourhood social mix and outcomes for social housing tenants: Rapid review. Centre for Research, Evaluation & Social Assessment & Superu. https://www.superu.govt.nz/social_housing_rapid_review#04-evidence-on-benefits-of-social-mix

Schaefer, K. (2013a). Die Wilde 13 [Documentary]. https://www.die-wilde-13.de/film/.

Schaefer, K. (2013b). Die wilde 13: Durch Raum und Zeit in Hamburg-Wilhelmsburg (2nd ed.). Geschichtswerkstatt Wilhelmsburg & Hafen.

Schäfer, S. (2017, March 10). Rassistische Diskriminierung? Mit türkischem Namen keine SAGA-Wohnung. MOPO.de. https://www.mopo.de/hamburg/rassistische-diskriminierung---mit-tuerkischem-namen-keine-saga-wohnung-26178322

Schepers, L. (2018). Leichen im Keller – Hamburgs koloniales Erbe. FINK.HAMBURG. https://fink.hamburg/snowball/hamburg-als-drehscheibe-des-kolonialismus

Schiffer, S. (2010). Rassismus in neuen und alten Medien. Institüt für Medienverantwortung. https://www.medienverantwortung.de/publikationen/videothek

Schiller, N. G., & Çağlar, A. (2013). Locating migrant pathways of economic emplacement: Thinking beyond the ethnic lens. Ethnicities, 13(4), 494–514.

Schlünzen, K. H., & Linde, M. (2014). Wilhelmsburg im Klimawandel – Ist-Situation und mögliche Veränderungen. [Berichte aus den KLIMZUG-NORD Modellgebieten, Band 4.]. TuTech Verlag.

Schmidt, T. (2012). Vulnerability through resilience? An example of the counterproductive effects of spatially related governance in Hamburg-Wilhelmsburg. Raumforschung Und Raumordnung, 70(4), 309–321.

Schönwälder, K. (2004). Why Germany's guestworkers were largely Europeans: The selective principles of post-war labour recruitment policy. Ethnic and Racial Studies, 27(2), 248–265.

Schönwälder, K., & Söhn, J. (2009). Immigrant settlement structures in Germany: General patterns and urban levels of concentration of major groups. Urban Studies, 46(7), 1439–1460.

Schultz, J., & Sieweke, J. (2008). Atlas IBA Hamburg: Wilhelmsburg neu vermessen. Verlagshaus Braun.

Shirlow, P., & Pain, R. (2003). The geographies and politics of fear. Capital & Class, 27(80), 15–26.

Silva, D. F. da. (2007). Toward a global idea of race. University of Minnesota Press.

Sinti-Verein Hamburg e. V. (2015). Sinti und Roma. Sinti-Verein Hamburg e. V. - Familienbildungszentrum Für Sinti Und Roma Zur Förderung von Kindern & Jugendlichen. https://www.sinti-verein.de/#uebersinti

Slater, T. (2006). The eviction of critical perspectives from gentrification research. International Journal of Urban and Regional Research, 30(4), 737–757.

Smith, D. p. (2004). Studentification: The gentrification factory? In R. Atkinson & G. Bridge (Eds.), Gentrification in a Global Context. Routledge.

Smith, L. T. (2012). Decolonizing methodologies: Research and indigenous peoples (2nd ed.). Zed Books.

Soric, M. (2020, June 22). Germany: Coronavirus exposes meat workers' plight. Deutsche Welle. https://www.dw.com/en/germany-coronavirus-exposes-meat-workers-plight/a-53902174

Sow, N. (2018, July 24). Sprachgebrauch in der Causa Özil: Über „Herkunftsdeutsche" (argh) und „Migrationshintergrund" (doppel-argh). Noah Sow Blog. https://www.noahsow.de/blog/sprachgebrauch-in-der-causa-oezil-ueber-herkunftsdeutsche-argh-und-migrationshintrergrund-doppel-argh/

Spiegel Online. (2012, December 28). Nach 124 Jahren: Hamburg löst seinen Freihafen auf. Spiegel Online. https://www.spiegel.de/wirtschaft/soziales/hamburg-loest-seinen-freihafen-nach-124-jahren-auf-a-874949.html

Spiegel Online. (2016, April 5). Hamburg-Blankenese: Anwohner von Nobelviertel blockieren Bau eines Flüchtlingsheims. Spiegel Online. https://www.spiegel.de/politik/deutschland/fluechtlingsheim-proteste-der-anwohner-in-hamburg-blankenese-a-1085577.html

Statistisches Amt für Hamburg und Schleswig-Holstein. (2006). Hamburger Stadtteil-Profile. Statistisches Amt für Hamburg und Schleswig-Holstein.

Statistisches Amt für Hamburg und Schleswig-Holstein. (2013). Ergebnisse der Mikrozensus-Zusatzerhebung in Hamburg 2010. Statistikamt Nord.

Statistisches Amt für Hamburg und Schleswig-Holstein. (2017). Lohn- und Einkommensteuerstatistik in Hamburg 2013: Grosse Unterschiede zwischen den Hamburger Stadtteilen (Spezial VIII/2017; Statistik Informiert...). Statistisches Amt für Hamburg und Schleswig-Holstein.

Statistisches Amt für Hamburg und Schleswig-Holstein. (2018). Hamburger Stadtteil Profile – Berichtsjahr 2016. Statistikamt Nord.

Statistisches Amt für Hamburg und Schleswig-Holstein. (2020). Hamburger Stadtteil-Profile: Berichtsjahr 2020. Statistikamt Nord.

Statistisches Landesamt der Freien und Hansestadt Hamburg. (2001). Wahlverhalten: Analyse der Wahlen zur Bürgerschaft und zu den Bezirksversammlungen am 23. September 2001 (No. 9; STATISTIK.MAGAZIN.HAMBURG). Statistisches Landesamt der Freien und Hansestadt Hamburg.

Stehle, M. (2006). Narrating the ghetto, narrating Europe: From Berlin, Kreuzberg to the banlieues of Paris. Westminster Papers in Communication and Culture, 3(3), 48–70.

Stewart-Harawira, M. (2013). Challenging knowledge capitalism: Indigenous research in the 21st century. Socialist Studies, 9(1), 39–51.

Stjernborg, V., Tesfahuney, M., & Wretstrand, A. (2015). The politics of fear, mobility, and media discourses: A case study of Malmö. Transfers, 5(1), 7–27.

Subventioniertes Wohnen: Studenten, ab in die Platte. (2013, April 1). Die Tageszeitung: taz. https://taz.de/Subventioniertes-Wohnen/!5070232/

Sudbury, J. (2007). Challenging penal dependency: Activist scholars and the antiprison movement. In J. Sudbury & M. Okazawa-Rey (Eds.), Activist scholarship: Antiracism, feminism, and social change (pp. 17–36). Paradigm.

SUPERURBAN PR und Stadt planbar. (2007). Weltquartier Wilhelmsburg: Ein Leben in der Kosmopolis. IBA Hamburg GmbH.

Tank, G. (2017). Gleich ist ungleich gleich ... In neue deutsche organisationen (Ed.), Gleich ≠ Gleich: Antidiskriminierungs und Gleichstellungsdaten im Gespräch.

taz. (2018, April 11). Racial Profiling auf St. Pauli: Polizeikontrollen kontrollieren. Die Tageszeitung: taz. https://taz.de/Racial-Profiling-auf-St-Pauli/!5494791/

Tedlock, B. (2005). The observation of participation and the emergence of public ethnography. In T. Denzin & T. Lincoln (Eds.), Handbook of Qualitative Research (pp. 467–481). Sage Publications.

Teelucksingh, C. (2007). Environmental racialization: Linking racialization to the environment in Canada. Local Environment, 12(6), 645–661.

The Association of the Ovaherero Genocide in the USA. (2017, January 27). Open letter to the city of Hamburg and the people of Hamburg. https://genocide-namibia.net/wp-content/uploads/2017/01/OpenLetter_OGA.pdf

Todzi, K. S. (2020, June 17). Koloniale Denkmäler: "Wir müssen uns damit befassen, wen wir ehren wollen." Zeit Online. https://www.zeit.de/hamburg/2020-06/koloniale-denkmaeler-denkmalsturz-restauration-kolonialismuskritik/komplettansicht

Todzi, K. S. (2018). Hamburg's first globalization. In J. Kokott & F. Takayanagi (Eds.), Erste Dinge. Rückblick für Ausblick/ First things. Looking back to look forward (pp. 166–170). Museum am Rothenbaum, Kulturen und Künste der Welt.

Todzi, K. S. (2017). Kolonialismus und Imperialismus. In Digitales Hamburg Geschichtsbuch. https://geschichtsbuch.hamburg.de/epochen/kolonialismus/

Tomiak, J. (2011). Indigenous self-determination, neoliberalization, and the right to the city: Rescaling Aboriginal governance in Ottawa and Winnipeg [Doctoral Dissertation]. Carleton University.

Trautwein, A. (2014a, May 9). Engagierte Wilhelmsburger. https://www.die-engagierten-wilhelmsburger.de/?tag=wilhelmsburg&paged=18

Trautwein, A. (2014b, July 4). Muslime wünschen sich Waschhaus in Finkenriek. Wilhelmsburg Online.de: Nachrichten für die Elbinsel. https://www.wilhelmsburgonline.de/2014/07/muslime-wuenschen-sich-waschhaus-in-finkenriek/

Tsianos, V. (2013). Urbane Paniken: Zur Entstehung des antimuslimischen Urbanismus. In D. Gürsel, Z. Çetin, & Allmende e.V. (Eds.), Wer MACHT Demo_kratie? (pp. 23–43). edition assemblage.

Twickel, C. (2011, March 25). Krisenviertel Hamburg-Wilhelmsburg: Einstürzende Neubauträume. Spiegel Online. https://www.spiegel.de/kultur/gesellschaft/krisenviertel-hamburg-wilhelmsburg-einstuerzende-neubautraeume-a-753127.html

Twickel, C. (2017, December 18). Hamburger Mietspiegel: Ausziehen? Besser nicht. Die Zeit. https://www.zeit.de/2017/52/hamburger-mietspiegel-wohnungen-kosten

Uitermark, J. (2014). Integration and control: The governing of urban marginality in Western Europe. International Journal of Urban and Regional Research, 38(4), 1418–1436.

UN Human Rights Council. (2007). Report of the Special Rapporteur on the right to education, Vernor Muñoz. Mission to Germany: 13-21 February 2006. United Nations General Assembly.

UN Human Rights Council. (2010). Report of the Special Rapporteur on contemporary forms of racism, racial discrimination, xenophobia and related intolerance: Githu Muigai, Mission to Germany. United Nations General Assembly.

UN Human Rights Council. (2017). Report of the Working Group of Experts on People of African Descent on its mission to Germany. United Nations General Assembly.

van Baar, H. (2017). Evictability and the biopolitical bordering of Europe. Antipode, 49(1), 212–230.

van Ham, M. (2012). Neighbourhood effects research at a crossroads: Ten challenges for future research. Environment and Planning A, 44, 2787–2793.

Vogelpohl, A., & Buchholz, T. (2017). Breaking with neoliberalization by restricting the housing market: Novel urban policies and the case of Hamburg. International Journal of Urban and Regional Research, 41(2), 266–281.

von Beust, O. (2005). Chancen einer Internationalen Bauausstellung für die Metropole Hamburg. In Sprung über die Elbe: Hamburg auf dem Weg zur internationalen Bauausstellung – IBA Hamburg 2013 (pp. 10–15). Freie und Hansestadt Hamburg, Behörde für Stadtentwicklung und Umwelt.

Waldretter. (2019). Waldretter – Lasst uns den wilden Wald retten! https://waldretter.de/

Walter, J. (2005). Entwicklung der Metropolen im Zeitalter der Globalisierung – Ausblicke auf eine Internationale Bauausstellung in Hamburg. In Freie und Hansestadt Hamburg, Behörde für Stadtentwicklung und Umwelt (Ed.), Sprung über die Elbe: Hamburg auf dem Weg zur internationalen Bauausstellung – IBA Hamburg 2013 (pp. 70–79). Freie und Hansestadt Hamburg.

Watt, P., & Smets, P. (Eds.). (2014). Mobilities and neighbourhood belonging in cities and suburbs. Springer.

Wehnelt, J. (2010, April 27). Musiker-Familie Weiss aus Hamburg: "Wir sind Zigeuner." Evangelisch.de. https://www.evangelisch.de/inhalte/99660/27-04-2010/musiker-familie-weiss-aus-hamburg-wir-sind-zigeuner

Wehrheim, J. (2018, June 6). "Soziale Durchmischung:" Lösung für soziale Probleme? "Soziale Durchmischung:" Lösung für soziale Probleme? Veranstaltungsreihe zur sozialen Gerechtigkeit in Hamburg, Haus der Patriotischen Gesellschaft, Hamburg.

Weiermann, S. (2020, June 11). Tod in Obhut der Polizei. Neues Deutschland. https://www.neues-deutschland.de/amp/artikel/1137740.tod-in-obhut-der-polizei.amp.html?__twitter_impression=true

Wenig, P. (2018, April 20). Der große Irrtum mit den Sozialwohnungen in Hamburg. Hamburger Abendblatt. https://www.abendblatt.de/nachrichten/article214070373/Das-Dilemma-mit-den-Sozialwohnungen.html

Wiedemann, J. (2018). Die Gartenfabrik: Wohnen und Produzieren auf der Elbinsel Wilhelmsburg in Hamburg [Thesis]. Wien.

Wildner, K., & Holtz, T. (2015). Zusammenklang: Sound of Wilhelmsburg – Eine auditive Stadtteilerkundung [SoundCloud]. Studierende des Fachs Kultur der Metropole der HafenCity Universität mit Prof. Dr. Kathrin Wildner und Tobias Holtz. https://soundcloud.com/soundofwilhelmsburg

Wolfe, p. (2006). Settler colonialism and the elimination of the native. Journal of Genocide Research, 8(4), 387–409.

Yeğenoğlu, M. (1998). Colonial fantasies: Towards a feminist reading of Orientalism. Cambridge University Press.

Yenirce, L. (2017a, May 19). Kolumne Inselstatus: Wer gentrifziert hier eigentlich wen? Die Tageszeitung: taz. https://taz.de/!5406331/

Yenirce, L. (2017b, July 17). Wie die Wilhelmsburger sich ihre ganz eigenen Regeln machen: Jenseits von Rauchverboten und Öffnungszeiten. Die Tageszeitung: taz. https://taz.de/!5426857/

Yiftachel, O. (n.d.). Displaceability: A southeastern perspective. Displacement Research and Action Network. Retrieved April 7, 2019, from https://mitdisplacement.org/symposium-oren-yiftachel

Yildiz, Y. (2009). Turkish girls, Allah's daughters, and the contemporary German subject: Itinerary of a figure. German Life and Letters, 62(4), 465–481.

Yuval-Davis, N. (1997). Gender and nation. SAGE Publications.

Zeit Online/dpa. (2019, August 14). Sozialer Wohnungsbau: Mehr als 42.000 Sozialwohnungen weniger in Deutschland. Die Zeit. https://www.zeit.de/politik/deutschland/2019-08/sozialer-wohungsbau-weniger-sozialwohnungen-deutschland

Zukunft Elbinsel Wilhelmsburg. (2012). Eine starke Insel mitten in der Stadt: Bürger-Engagement in Wilhelmsburg und auf der Veddel als Motor der Stadtteilentwicklung. Zukunft Elbinsel Wilhelmsburg e. V.

Zukunft Elbinsel Wilhelmsburg. (2014, January 14). Wilhelmsburger Inselpark: Jetzt doch Zaun statt öffentlicher Raum? https://zukunft-elbinsel.de/wilhelmsburger-inselpark-zaun-statt-oeffentlicher-raum/

Zukunft Elbinsel Wilhelmsburg. (2016, February 23). Grundrechte sind nicht verhandelbar und kein Thema für Abstimmungen! Zukunft Elbinsel Wilhelmsburg. https://zukunft-elbinsel.de/grundrechte-nicht-verhandelbar-kein-thema-fuer-abstimmungen/

Zukunftskonferenz Wilhelmsburg. (2002). Zukunftskonferenz Wilhelmsburg: Wilhelmsburg – Insel im Fluss – Brücken in die Zukunft [Bericht der Arbeitsgruppen Mai 2001 bis Januar 2002]. Zukunftskonferenz Wilhelmsburg.

Zwilling, J. (2012). Der IBA-Ansatz – Stadtteilentwicklung mit Schwächen. Idw – Informationsdienst Wissenschaft. https://idw-online.de/de/news505719

Social Sciences

Friederike Landau, Lucas Pohl, Nikolai Roskamm (eds.)
[Un]Grounding
Post-Foundational Geographies

2021, 348 p., pb., col. ill.
50,00 (DE), 978-3-8376-5073-0
E-Book:
PDF: 49,99 (DE), ISBN 978-3-8394-5073-4

Andreas de Bruin
Mindfulness and Meditation at University
10 Years of the Munich Model

2021, 216 p., pb.
25,00 (DE), 978-3-8376-5696-1
E-Book: available as free open access publication
PDF: ISBN 978-3-8394-5696-5

Gabriele Dietze, Julia Roth (eds.)
Right-Wing Populism and Gender
European Perspectives and Beyond

2020, 286 p., pb., ill.
35,00 (DE), 978-3-8376-4980-2
E-Book:
PDF: 34,99 (DE), ISBN 978-3-8394-4980-6

All print, e-book and open access versions of the titles in our list are available in our online shop www.transcript-publishing.com

CPSIA information can be obtained
at www.ICGtesting.com
Printed in the USA
JSHW052332260523
42316JS00005BA/58

9 783837 663877